M

EVANSTON PUBLIC LIBRARY

3 1192 01265 9007

508.78 Botki.D
Botkin, Daniel B.
Beyond the stony mountains

P9-DYE-633

DATE DUE

JAN 1 7 2005	
FEB 2 7 2005	
MAY 1 8 2005	
DEC 1 3 2005	

DEMCO, INC. 38-2931

SEP 2 4 2004

BEYOND THE STONY MOUNTAINS

BEYOND THE STONY MOUNTAINS

NATURE IN THE AMERICAN WEST
FROM LEWIS AND CLARK TO TODAY

DANIEL B. BOTKIN

OXFORD
UNIVERSITY PRESS

2004

EVANSTON PUBLIC LIBRARY
1703 ORRINGTON AVENUE
EVANSTON, ILLINOIS 60201

OXFORD
UNIVERSITY PRESS

Oxford   New York

Auckland   Bangkok   Buenos Aires   Cape Town   Chennai
Dar es Salaam   Delhi   Hong Kong   Istanbul   Karachi   Kolkata
Kuala Lumpur   Madrid   Melbourne   Mexico City   Mumbai   Nairobi
São Paulo   Shanghai   Taipei   Tokyo   Toronto

Copyright © 2004 by Daniel B. Botkin
Published by Oxford University Press, Inc.
198 Madison Avenue, New York, New York 10016
www.oup.com

Oxford is a registered trademark of Oxford University Press

All rights reserved. No part of this publication
may be reproduced, stored in a retrieval system, or transmitted,
in any form or by any means, electronic, mechanical,
photocopying, recording, or otherwise, without the prior
permission of Oxford University Press.

Library of Congress Cataloging-in-Publication Data
Botkin, Daniel B.
Beyond the stony mountains : nature in the American west
from Lewis and Clark to today / Daniel B. Botkin.
p. cm.
ISBN 0-19-516243-9
1. Natural history—West (U.S.)
2. Lewis and Clark Expedition (1804–1806)
I. Title.
QH104.5.W4 B67 2004
508.78—dc22   2003016664

*Book design and composition by Mark McGarry, Texas Type & Book Works, Inc.
Set in Minion*

9 8 7 6 5 4 3 2 1

Printed in China on acid-free paper

*To Jane O'Brian, one of the best naturalists I have known,
and who loved the Lewis and Clark trail, and loved travel
and adventure as did Lewis and Clark*

CONTENTS

ACKNOWLEDGMENTS

I am deeply indebted to Joan Melcher, who helped me structure this book, a difficult task arising from the attempt to combine then-and-now illustrations, as these were available, with a meaningful text. This book could not have been completed without her excellent and amazing help. I wish also to thank my wife, Diana, for her great help, as a professional copy editor—the best I have met—and for her interest in the subject, for encouraging me in my work on the book, and for her making life wonderful.

I thank Alex Philp for arranging a productive fall in Missoula, Montana, where many of my ideas were refined and broadened. He provided many kinds of support and remains one of the best critics of the ideas that are presented. He helped expand the work in content, in illustrations, and in the use of web-based technologies. James Peterson, former president of the Lewis and Clark Trail Heritage Foundation and a true "river rat" (in his own words), provided many photographs and information important to my work. Robert Zybach provided unique information about the history of forests in the Pacific Northwest. Ben Stout was a continual source of conversation and inspiration. Dick Pfilf, U.S. Forest Service retired, has kept my writing accurate and honest over the years.

Kirk Jensen, former executive editor at Oxford University Press, has been a colleague over many years and a consistent supporter of my writing. I thank him for the initial development of this book. Oxford editors Catherine Humphries and Elda Rotor, and Oxford researcher Cybele Tom provided excellent professional work to guide this book to publication.

And no twenty-first-century book about Lewis and Clark could be done without the excellent editing by Gary Moulton in his remarkable new edition of the journals. I thank him for that work, as well as for many fascinating conversations about the Lewis and Clark journey.

I have worked on, followed the path of, and written about Lewis and Clark for more than ten years, and I am indebted to many people I visited during my trips along the Missouri River and who guided me and provided invaluable information. These include Mark A. Brohman, Environmental Analyst Supervisor, Nebraska Game and Parks Commission, who took me on a wonderful full-day outing to see many of the restoration projects on the Missouri floodplain and was a continuing source of valuable information; Professor Thomas Bragg, University of Nebraska, Omaha, for sharing his insights about the Allwine Prairie and for discussions of the general status of prairies in Nebraska; J.C. Bryant, Refuge Manager (retired), Big Muddy National Wildlife Refuge, U.S. Fish and Wildlife Service, Columbia, Missouri, and Jim Milligan, Project Leader, Fishery Resources Office, Big Muddy National Wildlife Refuge, U.S. Fish and Wildlife Service, Columbia, Missouri; William Glenn Covington, Environmental Research Specialist, U.S. Army Corps of Engineers, Kansas City, Missouri, who spent a Sunday morning showing me Benedictine Bottoms and subsequently provided much helpful information about Kansas City and the Missouri River; Stephen R. Earl, Missouri River Project Engineer, U.S. Army Corps of Engineers, Omaha District, Operations Division, Omaha, Nebraska, who made it possible for me to travel on the boat *Mandan* for a day on the Missouri River; Gary Garabrandt, Chief Ranger, Fontenelle Forest Association, Fontenelle Forest Nature Center, Bellevue, Nebraska, who has taken me on many field trips over the years and has been a great source of information about the natural history of Nebraska and Iowa; Mimi Jackson, Director, Lewis and Clark Center, St. Charles, Missouri, for her stories about the floods on the Missouri and her kind help during my visit to the museum; Rob Leonard, Wildlife Management Biologist at Grand Pass Wildlife Area, Missouri Department of Conservation, Miami, Missouri, who took me on a tour of Grand Pass and explained the practices, policies, and ecology of the region, Dr. David Glenn-Lewin, Dean, Wichita State University, Fairmount College of Liberal Arts and Sciences, Wichita, Kansas, for many discussions about the prairie and for taking me on several field trips to see prairie lands, including Loess Hills and Ledges State Park; Gerald Mestl, Missouri River Program Manager, Fisheries Division, Nebraska Game and Parks Commission, who took Mark Brohman and me out on the river to visit Hamburg Bend and explained the ecological dynamics of the river; Larry Mason and his family at Tarbox Hollow Living Prairie, Dixon, Nebraska, for the tour of their bison ranch and helpful discussions about bison and prairies; Tom Motacek, Superintendent, Niobrara State Park, who was an informative and congenial host to a wonderful park; and Rick Plooster, Assistant Superintendent, Niobrara State Park, who took me on a beautiful boat ride on the Missouri River.

I thank Jane Weber, Director, Lewis and Clark Interpretative Center, USDA Forest Service, Great Falls, Montana; and Martin Erickson, Editor, *We Proceeded On*, of Great Falls, Montana, for helpful information and support for the idea of the book.

PREFACE

This book is about the natural history of the Lewis and Clark expedition as a guide to changes in nature in the American West from the beginning of the nineteenth century to today. As one of the best-documented explorations in the history of civilization, the Lewis and Clark expedition is a powerful aid to us as we try to understand what we have done to our surroundings. There are many common beliefs about what America was like before European exploration and settlement—part of our common knowledge—that are wrong. These misconceptions guide how we approach solving environmental problems and lead us astray. Only by knowing what the American landscape was like before we changed it can we decide how we would like it to be and how it could be "natural," whatever that word means. We have few windows on the American West—west of the Mississippi River. Here and there are remnants of "old-growth" and nature "preserves" that are *believed* to represent what that landscape was like. But nature *preserved* may be no more like the real nature than an Egyptian mummy is like the pharaoh in life. Lewis and Clark do not provide all the answers. Clark kept daily journals, Lewis wrote about whatever he believed important, and the three sergeants of the expedition were under orders to keep a journal and did. Thomas Jefferson, fascinated by natural history and science, instructed Lewis to observe nature in the American West and, prior to the journey, sent him to Philadelphia for crash courses in geology, botany, and zoology. Lewis and Clark, as experienced outdoorsmen, were good observers. However, they could not continually focus on the character of the countryside and its creatures. Their journals are remarkable in their detail and relative objectivity, especially given the severe circumstances of the expedition. As a professional ecologist who has done wilderness research, I know first-hand the difficulties of maintaining such records day after day. In the year 2000, several of us canoed downriver through the famous white cliffs region of the upper Missouri River. Historians, foresters, and scientists, we took with us copies of Karl Bodmer's famous paintings of the Missouri River route of Lewis and Clark and copies of Gary Moulton's wonderfully edited edition of the journals for that section of the expedition. Each day we worked hard—we thought—paddling in hundred degree temperatures downstream. We would pull up on the shore in the evening tired—too tired to keep our own journals—only to discover that Lewis and Clark and the members of the expedition had, each day in their time, traveled farther *upstream* than we had coasted down. And they were able to write in their journals faithfully each day.

Other scientists have noted that the records of Lewis are more useful than those of so-called professional scientists who wrote observations twenty years later. Some criticize Lewis's records simply because he was an "amateur," but that modern distinction had no meaning in his time—there were no professional environmental scientists, in the modern sense of being paid and making a living, back then. Keenness and objectivity of observation are the test of their work, not social standing. As an ecologist, I find their records strikingly good natural history observations.

Even so, the observations are by necessity somewhat spotty—not what ecologists today would call a scientific line-transect, with observations either at regular, specified intervals or at random points chosen from a table of random numbers. So we cannot expect to get a completely representative imprint of nature in the American West from their journals, but we can get invaluable insights of two kinds: 1) of nature itself; and 2) of how to survive, prevail, and live with nature, which Lewis and Clark did exceedingly well—more effectively, in my experience, than how we try to solve many environmental problems today. Therefore I believe this book will not only entertain—through its images of the past and present; through Lewis and Clark's colorful descriptions; and through the amazing tale of changes wrought in the land, by nature as well as by us, since they passed their way from St. Louis to the Pacific Coast—it will also inform the layman, the historian, the ecologist, and the environmentalist. It is with these hopes that I have taken on the task, as an ecologist, of writing this book.

Although there are many books about the expedition, few deal with the natural history of the American West as observed and experienced by Lewis and Clark. Most portrayals of the expedition offer a heroic story focused on the men (and one woman) as adventurers. But at the center of the story is two highly insightful, intelligent, and perceptive European men's contact with nature (and indigenous human cultures) in the American West. There appears to be growing interest in this central theme. And this theme has more than historical value. Lewis and Clark left St. Louis under the influence of European ideas and mythologies about nature: these ideas had captured the imagination of Thomas Jefferson. But they found themselves passing through eighteen ecological regions, each with its own challenges and wonders. If they had ignored the realities of these regions and marched forward under the prevailing beliefs about nature, they would probably have failed to reach the Pacific and failed to return to St. Louis.

Lewis and Clark's approach to the challenges and opportunities of the natural world serve as valuable guides to us today as we try to solve environmental problems. Ironically, although our modern information age is awash in data, we usually fail to measure the factors that are of crucial importance to solving environmental problems. If we do measure them, these measurements are rarely used: measurements are perceived through filters of established paradigms and cultural beliefs, which distort their interpretation and damage their utility. We believe when we should observe.

Lewis and Clark survived, prevailed, and came to understand themselves within nature in the American West. Theirs is a story that can appeal not only to Lewis and Clark buffs, but to all of us who want to understand how we can maintain and improve our civilization, nature, and the connection between the two.

It is common knowledge that the journey of Lewis and Clark was a fascinating epic: incredibly successful; full of adventures, near-disasters, and amazing coincidences; and replete with tales of courage and bravery. But it was more than that. It was a journey to discover the natural history of an unrecorded continent. As a result, it can be modern society's window on a nature we know little about but discuss often, believing that we do know it. On their way west, Lewis and Clark measured the distance they traveled; paced off the feet between river meanders; shot the sun with a sextant; looked at, touched, and tasted minerals; collected, described, and pressed new species of plants. They ate, wore, and wrote about wildlife. Their records tell us what nature was like before modern technology changed it: they have become a yardstick against which we can measure what we have done to the rivers and landscapes of midwestern and western North America.

Seeking to find the right route across the continent and to survive in the process, Lewis and Clark were not just keen observers, but also willing participants in an attempt to generalize successfully from a series of observations. It is a skill we are seldom taught and few of us learn: how to make reliable inferences from a selection of facts. More typically, we cannot believe that an event that we see in detail once may not be true in general. We fall into the unscientific trap of indefensible generalizations from too few observations. Lewis and Clark traveled up the river, but when they could, they strode on shore and climbed hills, bluffs, and mountains to get a view, to see a broader perspective. They measured and counted, they mapped and studied. In contrast, in our times—in this information age— we rarely obtain the information we most need about ourselves, our civilization, and our surroundings.

And so by experience, necessity, and Jefferson's plan, Lewis and Clark are our best external window on the reality of nature in the American West before it was altered by modern technological civilization. Their journey epitomizes our struggle to understand our effects on nature and nature's effect on us. Their journals provide clear and vivid insights into the past.

What is remarkable, and I believe unique, to the expedition of Lewis and Clark is that these two men took on the role of naturalist-recorders as seriously as they did their tasks of finding a route to help open up the West and making contact with and learning about Native Americans along the way. Human beings have long altered nature, but our knowledge of this is obscured by failed memories, confusion between myths and realities, and a loss of written historical accounts.

In preparation for the Lewis and Clark expedition, President Jefferson wrote to Meriwether Lewis that he should "record the mineral productions of every kind . . . Volcanic appearances . . . Climate, as characterized by the thermometer, by the proportion of rainy, cloudy, and clear days; by lightning, hail, snow, ice; by the access and recess of frost; by the winds prevailing at different seasons; the dates at which particular plants put forth or lose their flower or leaf; times of appearance of particular birds, reptiles, or insects." Through their historic journey, Lewis and Clark faithfully followed President Jefferson's instructions, recording the condition of rivers, prairies, forests, mountains, and wildlife, without romanticism, without ideology.

It is my hope that the material that follows will help the reader come to know the natural history of the American West both externally and internally, and that with this knowledge and appreciation we can move forward to a better use of our natural resources—for nature and for people.

BEYOND THE STONY MOUNTAINS

1

A PARTIALLY SETTLED LANDSCAPE

LEWIS AND CLARK NEAR ST. LOUIS

O N JANUARY 6, 1804, William Clark, in the midst of preparing for the journey west with Meriwether Lewis, was up and out at midnight to save one of his boats. "The banks were Caveing in," he wrote later that day, "and large Pees of the bank sliped in, which obliged all hands to, go Down & make all secure." He and Lewis had not yet begun their trip up the Missouri, over the Rocky Mountains, and down the Columbia, but nature was already challenging him. As he continued the preparations for the journey, the river rose and fell, requiring constant attention to the safety of the boats.

But in spite of the problems the weather and river imposed on him, and many other demands on him, from disciplining drunken soldiers to buying large quantities of supplies, he and Lewis kept notes about life around them. "The buds of the Spicewood appeared" on March 27, "and the tausels of the mail [male] Cotton wood were larger than a large Mulberry." In January they had begun recording the temperature, mounting a thermometer on a tree in the woods so that it would give the shade temperature—the right temperature to measure. "By two experiments made with Ferenheit's Thermometer," Lewis wrote in January, "I asscertained it's error to be 11°

too low." Not only did he measure and record, he checked out his measuring instruments.

While he and Lewis took time for such observations, sometimes their newly hired crew caused them trouble. Lewis wrote on March 3, 1804, in formal detachment orders, "The Commanding officer feels himself mortifyed and disappointed at the disorderly conduct of Reubin Fields, in refusing to mount guard when in the due roteen [routine] of duty he was regularly warned." And Clark noted in his journal on March 29, 1804, that "we have a trial of John Shields. John Colter & R Frasure which take up the greater part of the day," commenting in addition, "a violent wind from the N this mornig with rain, Some hail . . . river Continue to rise."

And so the days went by, and Lewis and Clark were already involved in what they would be doing for more than two years—leading a group of rough-and-ready men into unknown country, encountering challenging weather, rivers, mountains, and animals, yet always observing, recording, and measuring nature around them.

The thirty-three members of the expedition who made it to the Pacific coast and back were an interesting bunch, some better known than others. In addition to the captains, Meriwether Lewis and William Clark, Lewis assigned three men as sergeants—at first these were Charles Floyd, John Ordway, Nathaniel Pryor. Each was instructed to keep a journal and each did. With the death of Floyd, Patrick Gass became a sergeant and wrote his own journal which he published ahead of Lewis's. Other military members of the expedition were privates whose names were William Bratton, John Collins, John Coulter, Pierre Cruzatte, Joseph Field, Reuben Fields, Robert Frazer, George Gibson, Silas Goodrich, Hugh Hall, Thomas Proctor Howard, Francois LaBiche, Jean Baptiste LePage, Hugh McNeal, John Potts, George Shannon, John Shields, John B. Thompson, Peter M. Weiser, William Werner, Joseph Whitehouse, Alexander Hamilton Willard, and Richard Windsor. And then there were the non-military members, Toussaint Charbonneau, who was hired on at the Mandan Villages as an interpreter and was the husband of Sacagawea, the only woman on the expedition, and their son, Jean Baptiste Charbonneau. Also on the trip were Baptiste Deschamps, Pierre Dorion, and George Drouillard (also written Drewyer), the last the most important hunter and an interpreter, and finally, York, the only black member of the expedition and Clark's slave.

Meriwether Lewis and William Clark led an expedition from Illinois through parts of what are today Missouri, Kansas, Iowa, Nebraska, South and North Dakota, Montana, Idaho, Washington, and Oregon, traveling along the Missouri and Yellowstone rivers; over the Bitterroot Mountains of the Rockies; and down the Lochsa, Clearwater, Snake, and Columbia rivers. The trip lasted from May 14, 1804, to September 23, 1806, a total of 864 days (approximately two years and four months). From April 1805 through August 1806 "all communication with the world was suspended," and in this sense they were truly in the wilderness.

President Thomas Jefferson wrote to Lewis, "Record the mineral productions of every kind . . . Volcanic appearances . . . Climate, as characterized by the thermometer, by the proportion of rainy, cloudy, and clear days, by lightning, hail, snow, ice; by the access and recess of frost; by the winds prevailing at different seasons; the dates at which particular plants put forth or lose their flower or leaf; times of appearance of particular birds, reptiles, or insects." These instructions they followed, recording the condition of the rivers, prairies, forests, mountains, and wildlife, without romanticism, without ideology.

Lewis and Clark saw the land as people of European descent would never see it again, traveling by their own reckoning 4,134 miles on their outward journey and 3,555 miles by a shorter route on their return. It was America's greatest odyssey, beginning in St. Louis, navigating up the Missouri River and through the prairies into a winter with the Mandan Indians in North Dakota, traveling to the summit of the Rocky Mountains, and then down to the Columbia River, following that river down to the Pacific Ocean. Their journey has been called America's national epic of exploration, conceived by Thomas Jefferson, carried out by Lewis and Clark.

Before the journey began, from December 12, 1803, until May 14, 1804, they were camped at the mouth of the Dubois River just across the Mississippi River from St. Louis on the Illinois side. There they built a fortified camp—a group of buildings that they named for the river, Camp Dubois (literally Camp Woods). Here they planned the expedition, sought those with maps and knowledge of the western country, hired men, and gathered supplies. Clark described the landscape surrounding the camp: "The Country about the Mouth of Missouri is pleasent rich and partially Settled." The land east of the Mississippi—the Illinois side where the camp was— appeared "a leavel rich bottom" that was three miles wide, while the more distant uplands were "thinly timbered with Oake." The "Upper Country" along the lower Missouri River seemed "well Calculating for farming."

As Clark saw, they were in partially settled country—altered some but not very much by European civilization. Still, their living was rough by our standards. They built huts, but even Clark was not able to move into his until December 30, 1803. Camping outside, he wrote on December 22, that he saw "a verry great Sleat this morning, the river Coverd with running Ice," and that there was a "mist of rain, which prevents our doeing much to our huts to day." They fed on what they could hunt or bargain with from the Indians and the white settlers. The day before Christmas, Clark bought "a

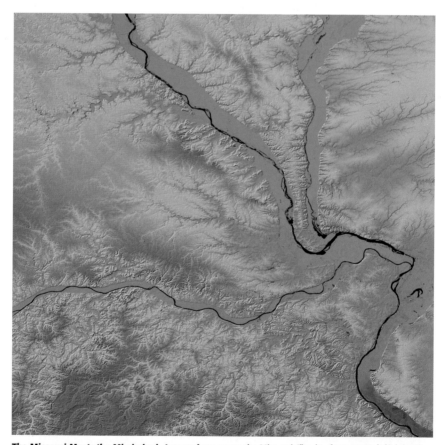

The Missouri Meets the Mississippi. As seen from space, the Missouri (flowing from upper left) joins the Illinois (flowing from upper right) and then joins the Mississippi (flowing from left across center). *Space-borne radar image, NASA Jet Propulsion Laboratory. Courtesy of NASA/JPL/NIMA.*

Cargo of Turnips" and wrote the day after Christmas that "one of my party Killed 7 Turkeys last night at roost." Game was still plentiful in this partially settled countryside, but so also were some European crops. What Clark bought to eat tells us the condition of the countryside, an insight, so to speak, from stomach to ecology. Clark observed, probably in May 1804, that "for 15 or 20 miles up each river, and about 2/3 of which is open leavel plains" he found fields of corn and wheat. "The Americans are Settled up the Mississippi for 56 miles as high up a[s] the Sandy river from thence across to the Missouries river," Clark noted, suggesting that even the partial settlement was restricted in area and that Camp Dubois was near the limits of it, at least to the north.

Looking across the confluence of the Missouri and Mississippi rivers today we see that the waters are so wide that the line of trees on the far shore looks no taller than the short hairs on a beaver's head and that there is still a mystique about the river. Here nature still looms large and powerful, even after major engineering alterations—channelization of the river, removal of snags and sandbars, construction of levees. One can see the Missouri emptying its churning, dangerous currents into the Mississippi channel. The flat landscape at the confluence of the Mississippi and Mis-

Modern St. Louis. *U.S. Army Corps of Engineers.*

souri, a view without distant vistas and obscured by vapors from the rivers, symbolizes the unknown that confronted Lewis and Clark at the beginning of their expedition and the myths and misconceptions about nature that they took with them.

Indian Mounds and Measurement

On Monday, January 9, 1804, William Clark, an outdoorsman, took time off from paperwork and other responsibilities and, enlisting John Collins, one of the men he had hired for the expedition, "went across a Prary to a 2nd Bank." There he came to a curious place. "I discovered an Indian Fortification," he wrote. Confronted with something new, Clark responded by taking measurements—a habit that would be his characteristic throughout the journey. "9 mouns forming a Circle," he recorded.

"[The base of] two of them is about 7 foot above the leavel of the plain on the edge of the first bank and 2 m from the woods," he continued. Looking around the mounds, he found "great quantities of Earthen ware & flints" and a "Grave on an Eminince." The mounds were not in use. They were ancient and abandoned. The local Indians knew little about them.

He had unwittingly stumbled onto what we now call Cahokia Mounds, the remains of the largest prehistoric earthen construction in the New World, built, according to discoveries by modern archeologists, between AD 700 and AD 1400. Looking back from our perspective, it was an ironic and curious discovery. He and Lewis were about to embark on a journey into what was perceived to be wilderness conforming to a formal set of European beliefs about nature in the New World, and Clark finds, quite accidentally and without any guidance, evidence that they were in the backyard of Native Americans who had affected that countryside much longer than anyone understood.

This suggested very different connections between people and nature than would be assumed from the European belief that this was "virgin territory." It demonstrated unequivocally that the lands along the Missouri River had long been settled by Native Americans who had carefully selected where to live in regard to nature's resources and who had had lasting effects on the countryside. Observing the connections between ancient peoples of North America and their environment, Clark found himself in a countryside that did not meet classical European assumptions about nature every-

Cahokia Mounds State Historic Site Today. *Cahokia Mounds State Historic Site.*

where—something he and Lewis would soon discover wherever they went. And just as the foundation of the largest mounds formed a base for others, the Indian cultures would provide a foundation for the expedition. The help that the Indians would give to the Corps of Discovery during their trip would be invaluable. It is fair to say that the expedition would not have succeeded without that help.

Today the mounds are preserved as the Cahokia Mounds State Historic Site. They form the largest prehistoric city in the New World north of Mexico. Approaching from St. Louis, entering the town of Collinsville, Illinois, one sees a tall mound rising surprisingly high above the level farmland, back behind fences in a large open field. Archeological studies indicate that, at its peak, the great mound city held about twenty thousand inhabitants, an urban population density similar to that of modern St. Louis. It is no accident that an ancient urban concentration and the modern city of St. Louis arose in the same area, because the environment near the confluence

Modern Setting of Cahokia Mounds. A few of the mounds, remnants of an ancient city, stand amid modern farmlands and lawns in a rural and suburban area of Illinois, east of St. Louis, Missouri. *D. B. Botkin.*

of the Missouri and Mississippi rivers offers many benefits. Just east of the confluence, the broad, flat floodplain offered good land for farming and habitat that attracted wildlife, drawn by the water and the vegetation that grew along the rivers. Rivers were then and remain important for transportation. And it is pleasant to be near a river.

The interactions and relationships between people and nature interested Clark, who, throughout the journey, looked at the countryside and envisioned future cities, towns, farms, forts, and harbors. And here he was, before the journey began, looking at Cahokia. The great mound city had been surrounded by a two-mile-long stockade of logs twenty feet high built from about twenty thousand trees. At the height of its development, Cahokia included 120 mounds, all of earth. A large ceremonial mound reached one hundred feet high, with a base of fourteen acres and a building on top more than one hundred feet long and forty-eight feet wide, and another mound was fifty feet high. More than fifty million cubic feet of earth were moved for the construction of the mounds.

The sophistication of the culture that developed here was revealed by archeological excavation in the 1960s that was spurred by a plan to put an interstate highway through the location. An archeologist, Warren Wittry, discovered oval-shaped pits the size of posts made from trees and arranged in arcs of circles. These appear to have served as celestial calenders, much as Stonehenge did in England, and have become known as Woodhenge. The posts mark the winter and summer solstices and spring and fall equinoxes. Like other early agricultural people, the Cahokians were dependent on the seasons for planting and harvest and needed a method to predict when changes would come.

A densely populated, defended city was possible here near the confluence of the Missouri and Mississippi rivers because the rivers and the surrounding countryside provided abundant natural resources. The floodplain's soil was frequently reenriched when the rivers flooded and deposited new soil carried from far upstream. Fish, freshwater shellfish, migrating and nesting water birds, and native mammals, including deer, were abundant in the complex habitats along the river and over the wide floodplain. The location of the largest prehistoric city in North America was not accidental but a direct result of geography.

Living close to the land and depending on it, the people of Cahokia responded to local differences in their natural resources. They farmed along the eastern floodplain of the Mississippi, but not along the western shore. At the time, the western shore was a series of bluffs and valleys, comparatively poor land to farm. Today, you would not notice this difference because the bluffs have been removed and the land leveled as part of the development of St. Louis.

The city of Cahokia began a gradual decline around AD 1300 and was abandoned by AD 1500. Nobody knows the fate of its people. The Indians Clark met near St. Louis knew no more than he did about the mounds. The history of the greatest city of ancient times in North America had been lost. Did the mound builders overuse their natural resources and then die off or migrate away because they had destroyed their local environment? Did a change of climate around AD 1400 make it impossible for them to live? Or did politics and war put an end to their culture? Nobody knows. Bones of the dead suggest some malnutrition and disease, so perhaps there was an environment-related decline in this civilization. Perhaps it is a history lesson we should pursue to see if there is any warning or message to help our civilization sustain itself and also sustain its natural resources.

Ancient European Ideas about Nature Confront
Realities of the New World

Clark's visit to Cahokia Mounds took place within an ecological region that we call the eastern deciduous forest. As Lewis and Clark traveled across the American West, they passed through eighteen ecological regions, each with its own challenges and wonders. These regions are eastern deciduous broad-leafed forest; tall-grass prairie; loess hills; buffalo country; short-grass prairie; Big Sky country—eastern Montana; Ice Age river pathway; upper Missouri River; Big Hole country; upper Clark Fork Basin; Bitterroot Mountains; Snake River system; Columbia River country; Cascades temperate rain forest; Columbia River estuary; Absaroka/Beartooth plateau; eastern Montana badlands; and the lower Yellowstone River. This book takes you through many of these, showing how the countryside in each has changed since Lewis and Clark passed through them. But Lewis and Clark's travels reveal more than scenery: They tell us about our ideas concerning nature and ourselves, now and in Jefferson's time. These ideas have not changed much, although the countrysides have changed greatly.

Lewis and Clark left St. Louis under the influence of European ideas and mythologies about nature; these had captured the imagination of Thomas Jefferson. The primary nature belief in eighteenth-century Europe was the balance-of-nature myth, which includes the following ideas: Nature, undisturbed by human influences, is constant over time and symmetric in space; there is a great chain of being, with a place for every creature, and every creature in its place, each performing a designed function; nature was designed and, if allowed to follow that design, functions perfectly, independent of human actions.

Constancy of nature was believed desirable and good—the best possible condition for all life. Geographic symmetry was believed necessary, desirable, and beautiful. The seventeenth-century bishop of Gloucester, Godfrey Goodman, echoed ideas of the Greeks and Romans about the necessity that nature be symetrical, writing that the height of the tallest mountain had to be the same distance above sea level as it was below it, because God created the natural world with a rule of porportion. But this was not the real North America, as Lewis and Clark were soon to discover.

The idea of nature's symmetry manifested itself in several ways in the planning of the Lewis and Clark expedition. Most important, Jefferson believed—like most people of his time—that the western mountains of North America must be symmetric with the Appalachians. Therefore, they would have the same width and height. This was an important aspect of Jefferson's plan to search for a water route to the Pacific. It suggested that the western mountains could be crossed in a day or so, without great difficulty, and that therefore an inland passage could be provided by the Missouri and Columbia rivers. This was believed to be not only possible, but *necessarily true*. The western mountains were "passable by Horse, Foot or Wagon in less than half a day," according to a late-eighteenth-century treatise promoting settlement of the West. One of the best maps of the late eighteenth century, by Aaron Arrowsmith—a map used by Lewis and Clark—said that the western mountains were "3520 Feet High above the Level of their Base." It is intriguing that one of the best mapmakers of the time accepted this belief enough to provide precise, but totally wrong, numbers.

In 1760 and 1761, Jefferson was taught by a Reverend Maury that the western rivers that flowed into the Pacific should reach as far east as the Missouri reached west—once again an expression of a necessary symmetry—and that the two rivers would be separated by a short and easy communication. Jefferson instructed Lewis to follow the Missouri River to its source, then cross the mountains and follow the Columbia River to the ocean. He did not suggest a more open-ended goal of finding the most practical route to the Pacific, because nature's symmetry had already dictated what that would be.

If the asymmetry of the mountains of North America had been known to Jefferson, or considered a likely possibility, the planning for the Lewis and Clark expedition might have been much different. First, Jefferson might have posed the goal differently: Seek the best water route to the Pacific. Second, he might have realized that a small party of men was unlikely to succeed in this endeavor, and he might have held off until the United States could, and was motivated to, fund a much larger military expedition.

Lewis was similarly educated and influenced by the predominant eighteenth-century belief in the balance of nature. But like Jefferson, he was also fascinated by natural history observations and by the potential of the recently developed scientific process. Thus he was, at least potentially, open to observations that might revise his ideas.

As experienced outdoorsmen, Lewis and Clark began their journey aware that they needed to observe their surroundings carefully if they hoped to survive in the wild.

But once on their way, they encountered an environment very different from the European perception. They faced two environmental challenges: the first to overcome the influence of dominant, but inappropriate, European ideas about nature; and the second to survive the heavy challenges of climate, topography, and wildlife largely unknown to western civilization.

Ironically, although we have progressed greatly in our technologies and many other sciences, our approach to the science of the environment lags and remains heavily influenced by the idea of the balance of nature. Two resulting ironies of our modern information age are, first, that we rarely measure what we need to know, and second, that if we do measure it, we seldom use the information. We would do better to follow the approaches of Lewis and Clark. Confronted—as we will see throughout this book— with the realities of nature in the New World, they observed, deduced, and responded in ways that were appropriate to what they experienced, although this required that they abandon the beliefs that formed the background of their European-based culture. In this way, their journey takes on a new and broad importance for us, beyond the fascination and excitement of the story of their expedition and beyond the confines of what we have come to call environmentalism.

Planning and Executing an Expedition

Lewis and Clark were great leaders and are seen today as traditional heroes of exploration. This is true, but there is more to leading a successful expedition than charismatic personality. A sobering thought about Lewis and Clark's success is that theirs was not the first but the *third* expedition sent out by Jefferson to find the best route west. The two before had simply disappeared, perhaps because their leaders lost interest or courage and turned away. Lewis and Clark succeeded not only because they were able to overcome the traditional European views of nature and deal with the nature that confronted them, on its own terms, but also because they planned carefully, beginning with Lewis's discussions with Jefferson, which led to Jefferson's sending Lewis to Philadelphia to take crash courses in botany, zoology, and geology so that he could identify, collect, and report on the natural resources he encountered.

The first remarkable thing about the conduct of the expedition is that the members planned for and kept careful and methodical records. It was the greatest wilderness trip ever recorded in pen on paper.

Another important aspect of the planning was the selection of the participants. Lewis began by telling Jefferson that he would lead the trip only if his good friend, William Clark, in whom he had great faith, accompanied him as co-leader. Lewis and Clark then selected the men for the expedition carefully. "Accept no soft-palmed gentlemen dazzled by dreams of high adventure," Lewis told Clark when they were interviewing people to make up the crew in 1803. "We must set our faces against all such applications and get rid of them on the best terms we can. They will not answer our purposes," he wrote.

Another aspect of good planning is choosing the right equipment, and one important part of good equipment is simply good maps. Much of Lewis's winter of 1803–1804 was spent in St. Louis talking with those who had traveled up the Missouri. Lewis copied and created maps as far as the land was known—from the mouth of the Missouri all the way to its middle portion, the present location of Bismarck, North Dakota. To put this in perspective, when the astronauts landed on the moon there were much better maps of the moon's surface than Lewis and Clark had available to them of the American West when they began their journey.

Good equipment is another quality of a successful expedition. Lewis was occupied for months with the selection, as well as invention, of devices to take on the journey. He purchased the best equipment available, including imported gunpowder, since at that time the quality of American-made powder was not good. He brought along the latest in technology, including a newly invented air rifle. He prepared for danger. In addition to the air rifle, Lewis brought three cannons mounted on the boats—not so much to win battles as to deter them—plus rifles, small flintlock pistols, muskets, blunderbusses, tomahawks, and scalping knives.

He purchased the best clothing available for the members of his expedition, including blankets and hooded coats and some clothing made from water-repellent material, as well as needles, awls, and thirty yards of flannel to make new clothing. He brought a wide array of tools, so that everything they needed could be repaired or made along the way. His medicine chest included the best available pharmaceuticals of the day. He also brought equipment, medals, beads, cloth, and other goods to give to or impress the Native Americans he would meet.

On the Missouri, the expedition—forty-eight men on the route from St. Louis to the Mandan villages, and then thirty-one men plus Sacagawea and her baby boy for the rest of the journey (the rest of the men, members of the U.S. Army, returned in the spring of 1805 to St. Louis with samples of animals, plants, and minerals)—traveled on a fifty-five-foot keelboat, called a bateau, and in two large wooden canoes, called pirogues. From St. Louis to the Mandan villages (near the site of modern Bismarck, North Dakota), the men traveled on horseback. The expedition left the Mandan villages with six canoes in addition to the two pirogues. Along the way, Lewis and Clark traded away horses, obtained canoes, traded *for* horses in order to cross the Bitterroot Mountains, and built canoes for the journey down the Columbia River system. An iron-framed canoe that Lewis had constructed especially for the trip was assembled after the group made its way past Great Falls. However, because they did not have the correct materials to seal the skins that covered the iron frame, the metal canoe sank. Wooden dugout canoes were built and used instead. In all, the equipment taken at the start of the trip weighed 3,500 pounds.

When appropriate equipment was lacking, Lewis invented new devices. A rifle he designed especially for the expedition became the model for the first mass-produced rifles of the U.S. Army. He packed carefully, making the most efficient use of the limited weight and space available to him. He designed watertight containers made of lead to hold gunpowder. When a container was emptied, it could be melted down to make bullets. He brought fifty-two of these containers, and they worked exceedingly well. Even his estimate of the number that would be needed turned out to be quite accurate. Twenty-seven were used on the outward journey, and only five of the remaining ones had cracked and allowed water to reach the powder by the time they were ready to return home.

How could Lewis have known that fifty-two canisters of gunpowder were enough? Experience in the field was part of the answer, and this, too, is an important factor in preparing for such an expedition. Both Lewis and Clark were experienced soldiers and outdoorsmen, accustomed to the hardships and the necessity for close observation in the countryside.

Beyond these were the inner human qualities of commitment, responsibility, and perseverance, qualities that, like boulders and water-saturated logs, are deep and dense but rarely come to the surface, which may be one of the reasons one becomes fascinated with these qualities in Lewis and Clark. And then there was just the willingness to do plain hard work, which Lewis, Clark, and the other members of the expedition showed repeatedly.

Finally, there are the elusive qualities of leadership, which the two men had in great abundance: an ability to command and lead, to make people want to go where they had not gone before and have them singing and dancing in the evenings, as they frequently did during the expedition.

Some of the writings in the early portions of the journals suggest that Lewis was not a tough outdoorsman and leader, but rather an urbane gentleman. When Lewis and Clark left Camp Dubois, Clark took the boats and sailed to St. Charles, just up the mouth of the Missouri. There, taking care of the last of the preparations, he waited for Lewis, who came overland from St. Louis. On his way from St. Louis to St. Charles, Lewis wrote on May 20, 1804, that the "morning was fair, and the weather pleaseant," and that he was "joined by Capt. Stoddard, Lieuts. Milford & Worrell together with Messrs. A. Chouteau, C. Gratiot, and many other respectable inhabitants of St. Louis." He bid "an affectionate adieu to my Hostis, that excellent woman the spouse of Mr. Peter Chouteau, and some of my fair friends of St. Louis," as if he were off on a spring picnic with the highest society of the town, not off into the wilderness of a continent. And he "arrived at half after six and joined Capt Clark, found the party in good health and sperits. suped this evening with Monsr. Charles Tayong a Spanish Ensign & late Commandant of St. Charles at an early hour retired to rest on board the barge" as if this were a well-organized tourist holiday for which he was just another passenger. But do not be fooled by this aside, nor underestimate Lewis's ability to move at several levels within society and without, for he could be both a gentleman of St. Louis and a rough-and-ready, experienced, thoughtful, curious, committed, knowledgeable, and tough leader of men.

Nature in the American West—including the great Missouri and Columbia rivers, the Bitterroot Mountains, and the prairies and forests—exists for us at two levels: internal and external. The first level is that of external knowledge: knowledge of natural resources, environmental issues, the names of animals, plants, and minerals, and understanding rational inferences about how the landscape and its life came to be and how they might be in the future. It is the level of detailed observation and records of natural history. The second level is that of feelings: how the countryside affects us and how we feel that we fit into that countryside. Like Lewis and Clark, we begin with the first, external level; these experiences lead us to the second.

2

CHANGING OLD RIVER

River take me along
In your sunshine sing me your song
Ever moving and winding and free
You rollin' old river
You changin' old river
Let's you and me river run down to the sea

—Bill Staines, "River"

Some people would think it was just a plain river running along in its bed at the same speed; but it ain't. The river runs crooked through the valley; and just the same way the channel runs crooked through the river.... The crookedness you can see ain't half the crookedness there is.

—C. D. Steward, "A Race on the Missouri,"
The Century Magazine, 1907

[The Missouri River] makes farming as fascinating as gambling. You never know whether you are going to harvest corn or catfish.

—Fitch, "A Race on the Missouri,"
The Century Magazine, 1907

The Missouri River challenged the expedition as soon as Clark led the crew by boat from Camp Dubois to St. Charles, Missouri. Clark left Camp Dubois on May 14, 1804, and the next day wrote, "the Boat run on Logs three times to day." The Missouri was wild, full of snags and sandbars even here, near its mouth, and the crew was just getting the equivalent of its sea legs. Clark observed that they hit the logs because the boat was "too heavyly loaded a Sturn." Ominously for their future travel up this river, he noted at the same time that the current was "excessively rapid, & Banks falling in." Such experiences were soon to become familiar.

Clark arrived at St. Charles with the boats and crew on May 16, 1804. He saw "a number Spectators french & Indians flocked to the bank to See the party." On May 20, Lewis joined them from St. Louis, and on Monday, May 21, Lewis and Clark and their group of forty-eight left St. Charles, Missouri, beginning their attempt to travel across the continent. They departed "at half passed three oClock under three Cheers from the gentlemen on the bank," Clark wrote. (Citizens of St. Charles claim their town was the starting point of the expedition as a whole, although Clark noted in his journal that Camp Dubois should be considered the expedition's point of departure.)

They were still in partially settled country, not the wilderness of the West that lay far ahead of them. In those first days of travel, they passed through a wooded landscape alongside the mighty Missouri River. Here, in eastern Missouri, the land was a mixture of the deciduous forests of the east and prairie patches characteristic of the land to the west. During this first stage of the journey, the biggest challenge was the Missouri River itself: The river's powerful currents endangered their boats, and its sandbars—some visible and forming a mazelike series of channels, others shallow, deep, or hidden just under the surface—stood ready to beach them. Its snags and sawyers (dead trees with branches bobbing in the current) could act like saws to cut through the side of any boat.

During the first two days of travel, the boats moved upriver without a major incident, but on Wednesday, May 23, Clark noted, they "Set out early run on a log: under water and Detained one hour." And the next day, they traveled through a reach of the river called "the Devils race grounds," and passing an island "wer verry near loseing our Boat in Toeing She Struck the Sands (which is continerly roaling) (& turned) the Violence of the Current was so great that the Toe roap Broke, the Boat turned Broadside, as the Current Washed the Sand from under her She wheeled & lodged on the bank below as often as three times, before we got her in Deep water." On the same day Clark observed that the bank on the port side was "falling in So fast that the evident danger obliged us to Cross between the Starbd. Side and a Sand bar in the middle of the river."

As the expedition moved up the Missouri River past the mouth of the Kansas River—the present-day location of Kansas City—Lewis and Clark and the men accompanying them found many beautiful areas along the shore and suggested some as fine locations for settlements or forts. On July 2, George Drouillard, a key member of the crew, their best hunter, traveled overland and told Clark that the countryside he had passed through that day and the day before on the south side "was generally Verry fine." That night, the expedition camped opposite an old Kansas Indian village, where there was a large island in the river and "extensive Prarie" beyond it. The island, Clark wrote, appeared to have "thrown the Current of the river against the place the Village formerly Stood" so that the current washed away the bank, forming an arc or natural harbor. He added that "the Situation appears to be a verry elligable one for a Town, the valley rich & extensive, with a Small Brook Meandring through it and one part of the bank affording yet a good Landing for Boats." The French had once located a fort here, he noted.

A River Leaves a Town

In 1837, Weston, Kansas, was established at this location as Clark suggested. Soldiers from nearby Fort Leavenworth saw the potential of the location, bought the land, and began to develop it. Weston's natural bay made a good port for boats to tie up, and there was fertile land nearby for farming; soil

An Island (left) in Today's Missouri River. Clark correctly understood how islands like this changed the river's current, forcing it to erode the nearby shore. *D. B. Botkin.*

that was dry and well-drained, suitable for buildings and their basements; and an ample water supply. The soldiers established a dock and built a main street that led away from the river. Settlers moved in quickly and set up a variety of shops and activities. The countryside was rich for farming. Tobacco farms were established and their products shipped downstream on boats that tied up at Weston's harbor.

A severe flood in 1844 damaged farmlands near Weston, and this was followed by an outbreak of diseases carried by water, such as typhoid. The town nonetheless continued to grow: by 1850 there were 5,000 residents. But in 1881 a bad flood occurred, and the Missouri cut a new main channel two and a half miles to the southwest of Weston. The river meandered away from the town, leaving Weston high and dry, with a harbor no longer at the foot of Main Street.

This event was natural for the river. It is natural for any river in a valley wide enough to allow meanderings, but the Missouri is especially famous for such meanderings. "Some people would think it was just a plain river running along in its bed at the same speed; but it ain't," a river man who raced boats on the Missouri River said a century after Lewis and Clark had traveled up it. "The river runs crooked through the valley; and just the same way the channel runs crooked through the river. . . . The crookedness you can see ain't half the crookedness there is."

A river on a wide and generally smooth floodplain does not flow in a straight line—or if it does, it does not maintain that straight line for long, especially if it is carrying a heavy load of sediment. Meanders are a natural form of a river, in part because the meander form keeps an even slope as the water flows downhill, minimizing the energy used by the river. In addition, even in a straight path, eventually chance occurrences will cause material to be deposited in one place and eroded in another—a log catches on the bottom, a pebble is pushed into the riverbed by the whirling water and catches hold. If the riverbed and its borders had been smooth, they are no longer. Because flowing water takes the path of least resistance, it begins to assume a sinuous shape around small obstacles, and the river starts to form a meander, creating shapes something like the reaction of spring steel that has been pulled straight and then released.

Although scientists can be sure that a river like the Missouri will meander, the exact location of any meander is influenced by chance events and cannot be predicted with complete accuracy. That is to say, the river wanders. It is also to say that the river is neither completely chaotic nor completely fixed.

A meander begins as a small bend in a river. Over time, the shape of a meander becomes more sharply arced, with more material deposited on the inside of the curve, where the river runs more slowly, than on the outside. The river erodes the outer, longer bank and deposits material along the shorter bank, nearer to the main channel. Eventually the meander takes on the shape of a near-circle, called an oxbow. A flood carries the waters across the short bank at the inside of the oxbow, cutting off the meander. This short channel becomes the path of the river, and a lake with the shape of a crescent moon remains, called an oxbow lake. Meanders of the Missouri have been measured to migrate across the floodplain at an average of about 250 feet a year.

This hydrologically natural event was a disaster for Weston. The fickleness of the Missouri led to a decline that almost destroyed the town. For years, Weston was a tiny village where little happened.

In the 1960s there was a renewed interest in Weston because of its historic buildings. The town began to redevelop as a tourist attraction and as a bedroom community for Kansas City, a short commute by today's standards. Today the population has risen to about fifteen hundred. Weston is nestled among the bluffs west of the Missouri River, a picturesque location. It is one of the prettier places for a traveler to stay along the Missouri River. If you walk down Main Street—only a few blocks, beginning in the hills and ending at what used to be the dock at the Missouri River—you can look across to the broad flatlands where once the main channel of the Missouri River flowed, now good bottomland for agriculture.

Learning the River

The crew gradually learned how to deal better with the river, which gave them plenty of practice. On June 9, 1804, once the men had had some time to learn about the river and work together handling the boats, the expedition passed Arrow Creek, near the modern town of Arrow Rock, Missouri, and just across the valley from what is now Overton, Missouri. Clark wrote that "the Sturn of the boat Struck a log which was not proceiveable." The current, he continued, quickly turned the boat "against Some drift &

Snags," which it hit "with great force. This was a disagreeable and Dangerous Situation, particularly as immense trees were Drifting down and we lay
imediately in their Course." Some of the men "leaped into the water Swam
ashore with a roap, and fixed themselves in Such Situations, that the boat
was off in a fiew minits." That they saved the boat, and did it quickly, was
remarkable and demonstrated an increasing skill, based on experience, in
dealing with the river. It was not the first nor the last of such incidents.

One of the worst stretches of the lower Missouri was near the location
of modern Camden Bend, Missouri. On Thursday, June 21, after several
days of hard rains, the river rose quickly, "3 inches last night," Clark
recorded, measuring as usual. Then he looked at the river and saw that an
island separated the flow and each side looked dangerous, "a most
unfavourable prospect of Swift water over roleing Sands which rored like
an immence falls." They used a long tow rope and the anchor—the Missouri flows with both speed and force that make it impossible to row or

**Sandbars and Meandering Channels in the Wild and Scenic, Unchannelized Lower Section of the Mis-
souri River.** This photograph shows the lower wild and scenic portion of the Missouri River, a section set
aside without channelization, levee construction, or other alteration of the water course (except, of course,
the effects of the dams upstream on water flow). Here, today, the lower Missouri River, flowing within its
wide floodplain, maintains the complex pattern of sandbars and channels that created difficulties for Lewis
and Clark. Ponca State Park, Nebraska, lies along this section of the river and is the best place for the visitor
without a boat to see the lower Missouri as it once was all the way to St. Louis. *James Peterson.*

pole against the currents. For a typical modern traveler, this would likely be a frightening experience, but Clark writes that they "got the Boat up with out any furthr dang. [damage] than Bracking a Cabin window & loseing Some oars." Just another day on the Missouri River. They were deeply imbedded with the river *as* nature and were accustomed to a life rugged in a way that is foreign to most of us. How did the Missouri appear to them? It is easy for us to transpose our feelings onto them—perhaps fear, terror, panic, a wish to be back home before a fire in a comfy chair—but this would be a mistake.

Meandering Meanders

On August 4, 1804, Lewis and Clark were a little north of the present location of DeSoto Bend National Wildlife Refuge, across the river from and not far north of what today would be Omaha, when once again the variableness and fickleness of the river became dangerously apparent. Clark wrote that the riverbanks were "washing away & trees falling in constantly for 1 mile." The next day the boats followed a large meander in the river upstream. In the evening Clark walked on the shore.

In "Pursueing Some Turkeys" he went downstream 370 yards on foot and found himself at the beginning of the meander, a distance he had measured to be twelve miles by river. Meanders not only lengthened the journey, they provided more of the dangerous, complex pathways for the expedition to thread the boats through. "In every bend the banks are falling in from the Current being thrown against those bends," Clark wrote, also noting that "agreeable to the Customary Changes of the river I Concld. that in two years the main Current of the river will pass through"—cut off the meander. Clark recognized the river's natural tendency to change its channel, to meander across its floodplain, to create sandbars and then erode them away, and to deposit soil on the edges and then undercut them into unstable cliffs.

Over the years, the meanders themselves migrate back and forth across the river valley. The Missouri River has wandered across the plains over thousands of years, eroding and depositing, like an artist working his oils over and over again on the canvas.

The Missouri as Landscape Painter

The geologist Brian Skinner has written that rocks are nature's books, and minerals its words. If so, then rivers are nature's landscape painters, brushing rocks and minerals, books and words, on the landscape. Rivers have a beginning: a young river cuts steeply through the book of rocks, creating cliffs. Rivers mature: they erode cliffs back into gentle hills; they create wide floodplains and meander through them. Life responds to this painted landscape. In the soils, microbes and plants read the words and push through the pages, abstracting life-giving nutrients. The river creates a landscape with flowing water, backwaters, side channels, stream-side zones, and uplands. Each is a different habitat to which a different collection of creatures has adapted. In the United States, one of the greatest of these painters of landscapes is the Missouri River. The Missouri picks its palette from the slopes of mountains and wind-formed hills. It carries these earth colors downstream and dabs the landscape with floodplains, terraces, and bluffs.

On this sculptured, painted landscape, Lewis and Clark pushed their small river crafts upstream, through the meanders, through the fallen sands, through the snags. They saw the river's sandy, silty painting at one moment in time. As mentioned earlier, a still-persisting belief of our age is that nature undisturbed by modern civilization is fixed, constant, steady, perhaps reliable and trustworthy. But the real Missouri River changed before Lewis and Clark passed its way, kept changing under their feet, and has changed since they left. The countryside, as a result, has also always been changing.

And from the time of the first steamboats on the Missouri River in 1818 until well into the mid-twentieth century, the variableness and unpredictability of the Missouri created difficulties for those who wanted to settle the land, farm it, build homes on its floodplain, and transport goods on the river.

The Missouri is one of Earth's twenty longest rivers, extending 2,315 miles from its origin at the confluence of the Jefferson, Gallatin, and Madison rivers in Montana to its mouth, where it meets the Mississippi at St. Louis. It flows north and then eastward from its origin, through Montana into North Dakota, then makes a big bend southward into South Dakota and down to Nebraska. From there, it flows southeast and east for a ways, forming part of the boundary between the two states, then turns south once again to form the boundaries between Iowa and Nebraska, Nebraska

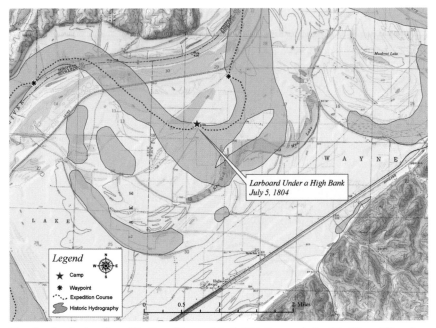

Map Showing the Location of the Expedition on July 5, 1804. The complexity of the Missouri River at the time of Lewis and Clark, with its meanders, oxbows, cut-off meanders forming oxbow lakes, and backwaters, is shown on this map of the expedition's July 5, 1804, campsite. They were near Independence, Missouri. © *Curators of the University of Missouri. Geographic Resources Center, Department of Geography, University of Missouri.*

and Missouri, and parts of Kansas and Missouri. Finally the river turns generally east and southeast, flowing through the state of Missouri to St. Louis.

This great river drains more than five hundred thousand square miles, or about one-sixth of the continental United States. The Missouri collects waters from the Bad River, the Blackwater, Cannonball, and Cheyenne rivers; and a series of rivers that flow into it from the south and west: the Gasconade, Grand, Heart, Judith, Kansas, and Knife rivers. It also picks up the waters from the northern plains extending into Canada—the Little Missouri, Moreau, Musselshell, Niobrara, Osage, Platte, Yellowstone, and White rivers—as well as the Big Sioux, Chariton, James, Little Platte, Marias, Milk, Vermillion, Sun, and Bad Teton rivers, which enter from the north and east.

Whatever enters this huge area of the United States comes out the Missouri. Drop a bottle with a message in it into a stream in eastern Montana or southern Canada north of Nebraska, and unless it rafts up on some sandbar or snag, it will float out at St. Louis. The Missouri begins as waters that fall on mountains fourteen thousand feet high in the Rockies, and it ends its journey only 400 feet above sea level. Starting in a major mountain range, the river flows in the western Great Plains through comparatively dry country that has been greatly altered by glaciers. The Great Plains give up their waters to the Missouri. In turn, the great river, with the help of vegetation, paints the surface into prairie.

The simplest way to understand the Missouri is to consider that it has four major geographic sections. The first extends from its headwaters nearly to Great Falls, Montana. The Madison, Gallatin, and Jefferson headwaters are in the Rockies, where rain and snowfall greatly exceed evaporation, and these rivers accumulate water and sediments that the Missouri River carries onto the plains. The second section extends from Great Falls to where the Milk River joins the Missouri near the Montana–North Dakota boundary. Here the river flows through semiarid plains along a geologically new pathway formed when Ice Age glaciers changed the Missouri from a river whose outlet was at Hudson Bay to one that flowed into the Mississippi. The third section extends from the Milk River—where the river joins its ancient bed, the bed of the pre–glacial era Missouri—to Yankton, South Dakota. Here evaporation exceeds rain and snowfall, and the river deposits sediments. In dry years, the river can lose water faster than it accumulates it from its tributaries. The fourth section is the last 825 miles from Yankton to St. Louis, where the river flows through a humid region of higher rainfall and low relief. Each section has its own scenery, its own hydrology, and its own characteristic, dominant species.

On the surface of our planet, the Missouri River acts as an irresistible force against which there is no immovable object. All earthly things that confront the Missouri, all that attempt to surround it, to seize it and hold it back, give way. If not now, then later. The mountains fall before it, as do the more meager works of mankind—levees, houses, and bridges.

While the Missouri's variableness was a great problem to Lewis and Clark, who were simply trying to travel it, its characteristics became worse problems to those of European descent who settled along its banks to farm in its floodplain; build houses, towns, bridges, roads, and railroads; and begin a settled and, they hoped, reliable, predictable life. This was not to be as long as the Missouri ran free, as farmers such as the Ryans (see

Handwritten on photo: *Missouri River, near T.J. Ryan Farm, Newcastle, Neb.* *3 51·28·*

Farmers at Ryan Farm in Newcastle, Nebraska, stare at the clifflike edge of a field that was partially washed away overnight by the Missouri River. Such variations caused by the river work against Western civilization's farming, which requires stabilized land. *U.S. Army Corps of Engineers.*

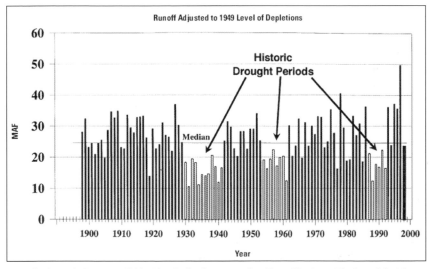

Annual Missouri River Runoff (the River's Flow) Measured at Sioux City, Iowa. The flow of the Missouri River varies from year to year, with a measured median of 25 million acre-feet a year, but with three major periods of drought in the twentieth century and several periods of high water that led to floods, including flows in the 1990s. *U.S. Army Corps of Engineers.*

illustration on page 27) learned in the twentieth century, when they witnessed the river wash away whole swatches of farmland overnight. Variation in flow from year to year was the "normal" characteristic of the Big Muddy, a changing, changeable river if there ever was one.

Attempts to Control the Missouri

> The river is the central fact [of one-sixth of the United States]. In its twisting and turning, ever easterly, from Three Forks to the Mississippi, the Missouri has succeeded in carving a crude but large question mark across the surface of one-sixth of the nation. This mark readily symbolizes the great array of problems which await satisfactory solution in the Basin.
>
> —MISSOURI BASIN SURVEY COMMISSION, 1953

Steamboats first ran on the Missouri on May 18, 1819, almost fifteen years to the day after Lewis and Clark started their trip. Steamboating as a dominant form of transportation lasted only four decades, ending before the Civil War and being replaced by railroads. Yet so treacherous was the Missouri that more than 450 steamboats were lost on the river during that time—a rate of about eleven wrecks a year. Some say that the average lifetime of a steamboat was four trips. Travel by steamboat up the Missouri became a major way west, especially important after the California Gold Rush and the Civil War, when immigration to the far west increased greatly. And so American society wanted a safer travel route. The government stepped in to try to help.

The government's first focus in modifying the great river was the removal of snags. These were notoriously dangerous to navigation—killers of steamboats since these craft began operating on the Missouri River, and destroyers of motorboats in the twentieth century.

From the beginning of the journey, the Missouri River made Lewis and Clark well aware of snags as a danger, as we have already seen. On July 5, when they were near Atchison, Kansas, Clark wrote that "the Boat turned three times" on some driftwood. And so there was an early movement to clear the river of these hazards. Congress first authorized the removal of snags from the Missouri River with an act passed in 1832. By 1838 more than two thousand large trees had been removed from the channel in the lower

Sunken Trees. Karl Bodmer, *Snags on the Missouri*, pencil and wash on paper (JAM 1986.49.129), Plate 141. *Joslyn Art Museum, Omaha, Nebraska.*

four hundred miles of the river. Between 1885 and 1910, these activities increased. In 1901, along 528 miles of the river, more than 17,000 snags were removed, along with more than 6,000 trees whose limbs hung over the river. The removal of snags was pretty much completed by 1950.

Next, the Army Corps of Engineers began a major program to dam and channelize much of the Missouri—building dams from the Montana–North Dakota border to just north of Sioux City, Iowa, and channelizing the river from Sioux City to St. Louis. This work began in the 1930s with the building of the Fort Peck Dam, a response to severe floods and droughts in 1930 and 1941. The Flood Control Act of 1944 authorized the Pick–Sloan Plan to construct five more dams. The dams were built to control water flow so as to improve navigation, prevent floods, and store water for crop irrigation during drought years. Then the Rivers and Harbors Act of 1945 authorized construction of a nine-foot-deep by three hundred-foot-wide channel from Sioux City to St. Louis.

The U.S. Army Corps of Engineers set out to remove the kinds of hazards that Lewis and Clark had observed—to make the river safe for people who lived and farmed on the floodplain; to provide a constant, reliable source of irrigation water from dams; and to make boat traffic safe and simple. There was a belief that barges would be a major way of transporting goods through the Midwest in the late twentieth century. But other forms of transportation—railroads, interstate highways, big trailer trucks, and airfreight—intervened, and the channelized Missouri never became a big moneymaker for the transportation industry. Today barges carry only three percent of the agricultural products of the region.

During the fifty years prior to the 1993 floods, twenty-five billion dollars had been spent on a system of levees, walls, and other flood-control measures on the Missouri-Mississippi river system by the Army Corps of Engineers and other state and private groups. Ironically, the vision these actions created of a calm and peaceful river led to some complacency: the greater the apparent control over the Missouri, the greater the faith people had in their own effectiveness, and the less alert they were to possible dangers.

Channelization and Construction of Levees on the Missouri at Indian Cave Bend Missouri. September 18, 1934, before construction begins. Water flowing behind levees slows so that it deposits sand along the shore and not in what is to become the main navigation channel. *U.S. Army Corps of Engineers.*

Construction Begins November 9, 1934. *U.S. Army Corps of Engineers.*

June 19, 1935. Sand Fills in Behind the Levees. *U.S. Army Corps of Engineers.*

May 23, 1946. The River Has Deposited Sediment. *U.S. Army Corps of Engineers.*

The Channelized Missouri River just four miles downstream from the unchannelized Missouri at Ponca State Park, part of the Wild and Scenic River system. Here the Missouri appears tamed, with straightened shorelines, simple smooth curves, no snags or visible sandbars. The farmed land in the floodplain appears stable, perhaps permanent. The landscape appears almost European in the sense of its geometric design and strong influence of human activities. This is the Missouri created by the Army Corps of Engineers to meet American society's goals for the river of the early twentieth century. Channelization of the river shortened it by 127 shoreline miles below the dams by cutting off meanders—a loss of about five percent of the length. *James Peterson.*

War of the Levees

There is only one river with a personality, habits, dissipations, and a sense
of humor. . . a river that goes traveling sidewise, that interferes in politics,
that rearranges geography, and dabbles in real estate; a river that plays hide
and seek with you today and tomorrow follows you around like a pet dog
with a dynamite cracker tied to its tail. . . . It cuts corners, runs around at
night, lunches on levees, and swallows islands and small villages for
dessert.

—G. FITCH, *THE AMERICAN MAGAZINE*, 1907

The city where Lewis and Clark met to begin their journey, St. Charles,
Missouri, experienced the effects of levees on the river nearly two centuries
later. Those who preceded Lewis and Clark surveyed, designed, and built St.
Charles, choosing its location with care. "The plain on which it stands,"
wrote Lewis on May 20, 1804, was "sufficiently elevated to secure it against
the annual inundations of the river, which usually happen in the month of
June." Indeed, St. Charles remained unflooded until 1993, when alterations
of the river by human actions made the town vulnerable to floods.

In recent years, this main street has become a private historical restoration—a series of lovely old houses now used as stores and bed-and-break-
fast hotels, catering to tourists. St. Charles is also home to the Lewis and
Clark Center, a small but excellent museum. In 1993—the year of the great
floods on the Missouri, the Mississippi, and the Red rivers—water began to
move up toward the museum building. The curators went out to watch the
water rise, expecting the worst, that the exhibits would soon be destroyed
by the floodwaters. But just before the water reached the edge of the
museum building, they heard a whoosh like the flushing of a toilet, and the
floodwaters quickly receded. They soon found out that a levee had given
way in Chesterfield, about ten miles distant, and that this allowed the
floodwaters to move into the countryside and away from St. Charles. This is
a clear example of how an alteration of one part of the river, the nearby levees, had effects on another, the waterfront of St. Charles.

Chesterfield responded to the failure of its levees by building bigger,
higher ones designed to resist a flood so great that it is likely to occur only
once in five hundred years. Some of the St. Charles businesses and some of
the government agencies near St. Charles propose doing the same, so that
there could be more development on the floodplain.

Lewis and Clark Museum at St. Charles, Missouri. *D. B. Botkin.*

The more the river is crowded in by taller levees, the faster it flows and the more erosive and dangerous it becomes. The result is a war of levees. A town or house with no levees or low levees suffers the flooding that is being avoided by the places with better protection. The alternative is to allow sections of the floodplain to flood naturally. Columbia Bottom, a large wetland near St. Louis being developed into a nature preserve from farmland, could be one of these areas. At Columbia Bottom there are more than four thousand acres to absorb floodwaters. Similar projects are found at other locations along the Missouri.

There is no better place than St. Charles to become directly acquainted with the dilemmas raised by a river that continuously changes its depth and is affected over its 2,300-plus miles by changes in seasons, snowfall, and rainfall—a dynamic and variable array of climate and geology. The question is, will we resist and fight against the forces of the river, or develop a regional design that allows for nature's continuous changes, providing some locations where floodwaters can spread out and, in a natural way, protect the towns along the river? This is a natural approach because the bottomlands have always been wetlands that flooded now and again, and they contain plants and animals adapted to these variations and able to recover from them.

The power of the Missouri's water was made clear in another way in 1993 to those who may have forgotten it or believed that the river was completely controlled. On Lisbon Bottoms and Jameson Island, near the remnants of the tiny hamlet of Lisbon, Missouri, a bridge had to be reconstructed after the flood. During that flood, the descending bank of a levee broke west of Glasgow, Missouri. Floodwaters took out about one mile of levee, one mile of Highway 240, and one mile of the GM&O railroad bed, creating a hole of several hundred acres, flooding thousands of acres of farmland, and cutting a pilot channel through a meander at Lisbon Bottoms.

Between a hundred and two hundred acres that were dry ground before the flood are now riverine habitat. The U.S. Army Corps of Engineers had put riprap (big rocks along the shore) there to stabilize the channel. But that flood came across Lisbon Bottoms with such force that it knocked the "nose" off Jameson Island. The levee that completely surrounded Lisbon Bottoms, meant to control the river, increased the impact. When the levee broke, Lisbon Bottoms filled up with water, and the water then blew the levee out on the lower side as it returned to the river. Local observers reported that it took a hundred dump trucks running twenty-four hours a day for two to three months to bring in enough rock to rebuild the levee, Highway 240, and the GM&O railroad.

The Six Great Dams on the Missouri River

In an average year, the water that flows down the Missouri River is enough to cover twenty-five million acres a foot deep—8.4 trillion gallons. The average water use in the United States is one hundred gallons a day per person—high compared with the rest of the world. In some countries, people make do with ten gallons or less a day. At one hundred gallons a day per-person, the Missouri's flow is enough to provide domestic and public water in the United States for about 230 million people. With a little water conservation and a reduction in per-capita use, the Missouri could provide enough water for all the United States, so great is its flow.

The six major dams on the river were designed for several purposes: to hold back and control floodwaters, to release water so that there would always be enough in the channels for safe navigation, to keep enough water in the reservoirs to provide that flow in years of drought, and to provide water for irrigation.

There are two kinds of dams on the Missouri: big storage dams and control dams. The storage dams are the ones farthest upstream: Fort Peck, Garrison, and Oahe. Each can store approximately twenty-five million acre-feet, and together they store a three-year supply of Missouri River water flow even if there were no rain or snow.

"Under perfect conditions, the storage drops to fifty million acre-feet—a two-year supply—in March, just before spring runoff from the mountains," an Army Corps engineer has explained. "Then we hope that the spring runoff will just be enough to fill the reservoirs back up to a three-year supply." As the upper reservoirs fill, water is released to the three lower dams, which then release water as needed so that the channel stays as close as possible to desired constant conditions.

"But when the weather doesn't cooperate, somebody is bound to be unhappy. If there is a drought then the storage may fall below two years supply," another engineer continued. "If it is a very wet year then the dams reach their capacity and water has to be released, with flooding the result. One of the things nobody planned on originally is that a lot of recreation grew up on the reservoirs. Now in a drought year, when we have to let the water level fall in the dams, a lot of people complain that we are ruining the recreation. People come from all over the West now to fish in the reservoirs. Upstream people have become used to the reservoirs and want them at a high level. Downstream they want no floods. The farmers want land that is farmable.... We can't solve everybody's problem at the same time."

Construction of the dams also meant that large areas of land would be covered by the reservoirs and lost as fish and wildlife habitats. The big three of the six dams impound almost a million acres: Fort Peck, 249,000 acres; Garrison, 368,000; Oahe, 371,000. The three downstream, smaller dams, impound something under 200,000 acres: Big Bend, 61,000; Fort Randall, 102,000; Gavins Point, 32,000.

Today, the idea of altering so much of a major river seems strange to many of us, but during the 1920s and 1930s, with the Dust Bowl and the

Depression, our society embraced the idea that we needed big dams on our big rivers to provide water for irrigation and electricity for power. During the same era in which the big dams were being built on the Missouri, dams were also being built on the Columbia, the other great river of the Lewis and Clark expedition. Woody Guthrie was one of the first employees of the Bonneville Power Administration, set up to build the Columbia River dams. He was hired to write songs about the huge projects and to popularize dams for irrigation and power, and he believed in the cause.

"Roll on, Columbia, roll on, roll on, Columbia, roll on. Your powers are turning the darkness to light, so roll on, Columbia, roll on," Woody wrote. The Grand Coulee, he wrote, was "the biggest thing ever made by a man, to power our factories and water our land, so roll on, Columbia, roll on."

Woody Guthrie was a social activist, a union organizer, a political radical; his support of these projects and his songs about their benefits show how different our society's attitude was about the rivers. They were just "a thousand years of water going to waste," he wrote in another song.

To understand many of the dilemmas that face our society, one has to understand the social context within which we attempt to solve environmental problems. In a time of desperation for many people, using the power of the rivers to create jobs and better the lives of the impoverished was seen as a social good and an important political movement.

The irony from our perspective is that these dreams of social good came at a high environmental cost. Not only were large areas of floodplain habitat lost, but controlling the flow did away with seasonal patterns of change. Before channelization and control of the flow, there was a natural hydrological seasonal pattern, with two floods in the spring. The first was in March, when the ice melted on the river and snow melted on the plains, as Lewis and Clark saw during their winter with the Mandans. The second came in June, when the snow melted in the Rockies and there was rainfall in the river basin. Usually the June flood was higher. Fish and wildlife had adapted to these seasonal variations, some requiring it as part of their life cycle. So the elimination of the seasonal variations threatened fish and wildlife. This only became a major public concern with the rise of environmentalism, beginning in the 1960s, when the finishing touches were taking place on the Army Corps channelization and dams on the Missouri River.

Straightening the Big Muddy

The channelization of the Missouri River led to ironic consequences, illustrated at many locations on the river. At DeSoto Bend National Wildlife Refuge, the U.S. Army Corps of Engineers constructed a channel that cut through a meander to shorten river travel by seven miles, avoiding the DeSoto Bend. Beginning work in 1960, they built levees to cut off the meander, forming an oxbow. They did this even though oxbow lakes were continually being formed by the Missouri through natural processes. These lakes are scattered over the countryside, and many are recreational parks, such as the Lewis and Clark State Park in Iowa.

Before the great floods of the Missouri in 1993 and 1996, it seemed that channelization of the river had tamed the wonderful wild Missouri of fact and folklore, turning it into a placid stream. We had made a Faustian bargain with the river, gaining short-term stability—a chance to build and live on the floodplain, and to farm that floodplain for a number of years without worrying about dreadful floods—and in exchange losing the renewing sediments that created the fertile farmland in the first place and courting rarer but more dangerous floods in the future.

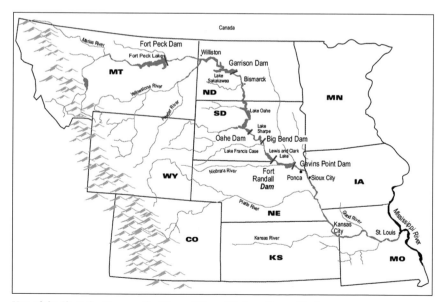

Map of the Six Major Dams on the Missouri River. *U.S. Army Corps of Engineers.*

DeSoto Bend National Wildlife Refuge is typical of the channelized and controlled Missouri River. Before the floods of the 1990s, low wetlands back from the river often had large willows and cottonwoods, species characteristic of those habitats. But the willows were much larger—probably much older—than was typical of the forest at the time of Lewis and Clark. There was also a dense understory of flowering dogwood, unlikely to be seen along an undammed and unchannelized river subject to seasonal floods of much smaller magnitude than the major 1990s floods, because dogwood cannot withstand flooding and the floods would bulldoze the small trees away.

At the time of Lewis and Clark, floodplain forests would have had a "cathedral" look—tall, arching trees with little understory. Many dead logs would have lain along the bottomland at the time of Lewis and Clark, some carried down by the river, the rest from trees that had fallen and remained in place. Some kinds of trees adapted to those wet, frequently flooding habitats—such as ash and elm—were common then on floodplains. Elm has declined both because of habitat destruction and because of the introduced Dutch elm disease, so that one of the few places you can still find an occasional elm growing wild today is in such floodplains, where the sparseness of elms makes spread of the disease slow or unlikely.

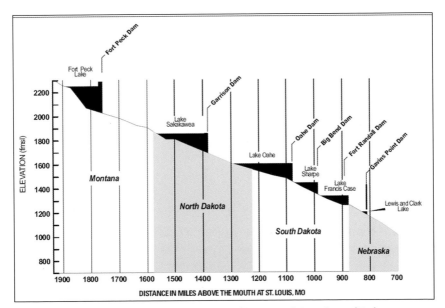

The Six Big Dams on the Missouri, Their Elevation and Storage. *U.S. Army Corps of Engineers.*

After the 1993 flood, well-intentioned works of human beings on the river lay in disarray. Neat, straight banks built by the Army Corps of Engineers were gone, washed away; the even line of boulders a jumble of rocks.

Since the time of Lewis and Clark, the Missouri River has been teaching the same lessons, but rarely have we listened, and thus rarely have we learned. We thought that our mechanized projects were a rational approach to the river, but it hasn't worked out that way. There is a rational approach we can take to living with the river and benefiting from its waters, and that is to conserve its living resources; enable it to fertilize and restore the land; accept and allow for its variations, allowing for them, by providing places for floodwaters to spread; and understand that such a river is partially but not completely controllable. We must understand that a complex landscape, a mosaic of many kinds of habitats, is important to fish, wildlife, and vegetation—and therefore to us. The way to save the river and the life that

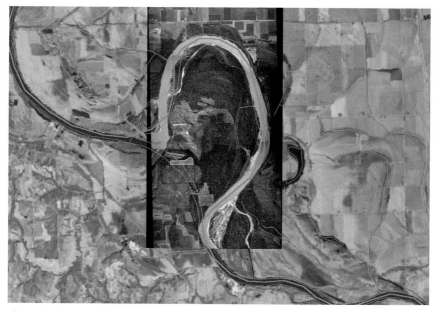

Changes in the Missouri River at DeSoto Bend: Landsat 5 and Black and White Historical Aerial Photo with Lewis and Clark Trail Line. A blend of images from satellite shows changes in the Missouri River near De Soto National Wildlife Refuge north of Omaha. The dark blue band is the route of the Missouri River before channelization. The light gray-blue arc is the oxbow lake formed when the meander at De Soto National Wildlife Refuge was cut off by a man-made channel. The yellow lines shows the path of Lewis and Clark expedition. *GCS Research.*

depends on it, and to best protect man-made structures is to build *with* the wildness, letting the river run free as much as we can. We need to move past the outdated idea that a river can be managed like a household plumbing system and accept its true ecological and geological character that of the landscape painter meandering over the countryside.

3

A COUNTRYSIDE PLEASANT, RICH, AND PARTIALLY SETTLED

THROUGH THE EASTERN WOODLANDS

DURING THE FIRST PART of the journey—from St. Charles to the Kansas River, in what is now the state of Missouri—Lewis and Clark's eyes were mainly on the river, not on the land. The river's challenges were enough to occupy all of the men. The land was primarily of interest as a source of wildlife for food—deer, turkey, and bear. Lewis and Clark often did, however, go onto the land, and in some cases they commented about the landscape.

On Saturday, June 2, 1804, the expedition reached the Osage River (near modern Osage City), where Lewis used the sextant to measure longitude. Clark measured the height of the land above the river and the distance between the Osage and the Missouri rivers. They climbed a limestone rock where, Clark wrote, he "had a delightfull prospect of the Missouries up & down, also the Osage R up."

The land, like the river, challenged the men. Clark notes that two of their crew, George Drouillard and John Shields, whom they had sent with the horses, met them that evening after a week of living on what they could hunt, were "much worsted" from spending "the greater part of the time [in] rain." Also "they were obliged to raft or Swim many Creeks." But these men were tough, as were Lewis and Clark, and reported, in spite of their personal conditions, "a flattering account of the Countrey" where they found "som fine Springs & Streams." On the river or on the land, the men of this expedition traveled in what we might consider great difficulty. For most of us, simply passing through that countryside would be more than enough, but these two, like Lewis and Clark, were able to observe and report on the condition of the land and its potential uses.

Limestone is the dominant rock along the lower Missouri, and even today, because limestone is a base for fertile soils, it creates a pretty landscape. From a modern highway this is rarely visible, but by canoe on the lower Missouri one sees a green, rich landscape, surprisingly similar to what Lewis and Clark would have seen.

This forested countryside would have been familiar to Lewis and Clark, as they were within the ecological region called the eastern deciduous forest, the same region as their homes in Virginia and the states they had traveled on their way to St. Louis. The eastern deciduous forest develops where rain falls more than twenty inches a year, and typically thirty-five inches or more; where precipitation is fairly evenly distributed throughout the year; where summers are hot and humid; and and where winter temperatures drop considerably below freezing. (The trees of this forest type are so adapted to a cold climate that most require frost and freezing conditions to reproduce and have their seeds germinate and survive.) As they traveled west through what is now the state of Missouri, the climate became drier, and they entered a transition between the eastern deciduous forest and the tall-grass prairie. As they neared the Kansas River, the prairie encroached more and more on the forest, and the forested land became restricted to a belt along the water's course.

Missouri, not yet a state, was a mixture of new clearings and forests of oaks, hickories, and other trees on uplands, with elms, cottonwoods, and willows found along the Missouri River and other streams. The partial settlement consisted of farms, scattered European-style villages, and Indian settlements, increasingly affected by the newcomers.

To Clark's eye, the countryside where they set out appeared fertile land good for farming. Probably in May, he wrote that "in the point the Bottom is extensive and emencly rich for 15 or 20 miles up each river, and about 2/3 of which is open leavel plains in which the inhabtents of St. Charles &

potage de Scioux [probably the portage of the Sioux] had ther crops of corn & wheat. on the upland is a fine farming country partially timbered for Some distance back."

Arrow Rock and Cottonwoods

Woodlands were common along the river, and as Lewis and Clark moved westward into ever drier country, prairie occupied more and more of the landscape and trees were concentrated nearer to the river. While the Missouri River was treacherous, it also created a well-watered and fertile bottomland where trees that were adapted to these conditions grew in great abundance. On June 9, Clark wrote that they passed a place called "Prariee of Arrows," which was below a bluff called "Arrow Rock" and where a stream called the "Creek of Arrows" flowed into the Missouri. Clark observed a "Delghtfull land" on the south side of the river. The next day he noted that they "passed a part of the River that the banks are falling in take-

Cottonwoods Grow Rapidly after the 1993 Missouri River Flood. The river covered crops with new sediment, and in five years cottonwood trees had grown to 30 feet high in stands so dense it was almost impossible to walk through them. These cottonwoods appear in this view from the town of Arrow Rock as the light gray green beyond the nearest stream. *D. B. Botkin.*

ing with them large trees of Cotton woods which is the Common groth in the Bottoms Subject to the flud."

The bottomlands were full of cottonwoods, which came and went at the will of the river, sprouting after a flood, dying when the current cut the soil out from under them. These same natural processes are working today at the same location. After the expedition, Arrow Rock developed as a crossing point on the Missouri River because Arrow Creek provided a way up and over the steep bluffs. The town of Arrow Rock blossomed briefly. The Santa Fe Trail, which originates in Independence, Missouri, came this way.

Today Arrow Rock, Missouri, is a small village of about seventy residents and is registered as a National Historic Landmark, with a main street of pleasant, well-maintained houses and shade trees. Arrow Rock State Park stands on the edge of a bluff, providing a magnificent view of the Missouri River. It has a large, manicured lawn of playing fields and a few trees—oaks and other upland species. Below flows the Creek of Arrows, much as Clark had described it.

Beyond the creek and far below is a vast field of cottonwoods, and beyond the first field of trees is a wide expanse of bare soil and another field of cottonwoods mixed with willows. These cottonwoods sprouted on former agricultural lands after the 1993 flood, creating stands so dense they were almost impossible to walk through half a decade later. The trees grew remarkably fast: some were twenty and thirty feet high and up to five inches in diameter— as thick as a wrist—after five years. Under good conditions, trees of the upland forests of Missouri and the states to the east— like the oaks growing on the top of the bluff in the woods bordering Arrow Rock State Park—grow a foot or maybe two feet taller and a half-inch wider a year. In a poor year, upland trees, especially those adapted to the deep shade of older forests, grow hardly at all, just enough so you can see a growth ring when you look at the stump. Some trees, including some of the oaks, can persist for years with this kind of minimal growth.

Cottonwood is a fast-growing tree that is characteristic of what ecologists call the early successional stage in the development of forest—the time soon after a disturbance such as the 1993 flood, when light, water, and nutrients are in great abundance and there is little competition. Willows are much the same, germinating and sprouting on newly formed floodplain soils. Sometimes the willows come in first, especially on the coarser soils of sandbars, followed by cottonwoods. Sometimes, especially in heavier soils, cottonwoods sprout immediately. Cottonwood is also the dominant species in what appear to be mature bottomland forests on the Missouri River.

Lewis recognized this process of natural change in dominant species. Reflecting on the vegetation of the Missouri River during the winter, Lewis observed that "when these rivers [Missouri and its tributaries] form new lands on their borders or islands in their streams, which they are perpetually doing, the sweet willow is the first tree or shrub which usually makes its appearance, this continues one two or three years and is then supplanted by the Cotton wood which invariably succeeds it."

Lewis also observed that, once the trees were rooted, they helped build up the floodplain soil. "The points of land which are forming all ways become eddies when overflown in high water," he wrote. "These willows obstruct the force of water and makes it more still which causes the mud and sand to be deposited in greater quantities." Lewis added that "the willow is not attal imbarrased or injured by this inundation, but (the moment

Doomed Cottonwoods on the Missouri River near Vermillion, South Dakota. The Missouri River continuously deposits sediment, forming new land where cottonwoods and willows sprout and grow, only to be toppled years later, as the Missouri erodes the land it deposited, taking back what it once gave. In this photograph, the river is undermining the shore and cottonwoods are beginning to topple. This scene is within the tall-grass prairie region where cottonwoods are among the few tree species and are restricted to land along flowing waters. The same process of death and renewal, however, occurs wherever trees can grow near the river and is characteristic of the eastern deciduous forest, the first ecological stage in Lewis and Clark's journey. *James Peterson.*

the water subsides) puts forth an innumerable quantity of small fibrous roots from every part of its trunk near the surface of the water which further serve to collect the mud."

He also observed that the trees thinned over time. "As the willow increases in size and the land get higher (and more dry) by the annul inundations of the river, the weeker plants decline dye and give place to the cotton-wood which is it's ordinary successor, and these last in their turn also thin themselves as they become larger in a similar manner and leave the ground open for the admission of other forest trees and under brush."

This self-thinning is happening today on Jameson Island near Arrow Creek. The faster-growing trees shade out the slower-growing ones, and the slower-growing ones die. The organic material from these dead trees improves the already-rich soil. The cottonwoods and willows are a clear example of life's great productivity and its ability to restore itself when conditions are right.

Floodplain forests have diminished greatly since the time of Lewis and Clark. These forests covered about seventy-five percent of the Missouri River floodplain in 1826, but only thirteen percent by 1972. The clearing of land for farming was a major cause of this decrease: between 1826 and 1972, cropland increased from eighteen percent to cover eighty-three percent of

An Engraving from 1915 Shows Firewood Collected for Steamboat Fuel. From an engraving on lithographic limestone quarried in Floyd County and published in Clement Webster's 1915 issue of *Contributions to Science* to illustrate the high quality of this Iowa stone for printing. *From IOWA - Portrait of the Land, Iowa Department of Natural Resources, Iowa, 2000. Adapted from Clement Webster's 1915 issue of* Contributions to Science. *p. 22.*

the floodplain. But channelization also had its effect: sprouting and survival of cottonwoods and willows declined greatly because of the reduction in spring flooding. The habitat had been changed and was no longer available or suitable to these trees.

The view from Arrow Rock State Park illustrates a general rule about life: Most species do well as long as their habitats are in good condition. It is better to have a small population in a good habitat than a large population in a poor habitat. And here the habitat for cottonwoods is nearly perfect—a floodplain whose soil was just renewed, laid bare of other vegetation so that cottonwood sprouts could thrive.

The scenery at Arrow Rock today is not an exact duplicate of what Clark saw, in part because of natural changes in the river channel and in part because of loss of backwaters, due largely to human channelization of the river. But the land is undergoing the same ecological processes he and Lewis observed. Cottonwoods grow in abundance after disturbance to the bottomlands, just as Lewis and Clark described them doing almost two hundred years ago, and today the species of trees change with the elevation above the river, just as Clark said. These patterns and processes are gateways to opportunities to restore the landscape. Given a good habitat, a small number of mature cottonwoods can provide the seeds to fill a valley.

The expedition came across caves, common where limestone is the primary bedrock, as it is in this part of Missouri. The limestone has been valuable enough to mine, and one quarry is readily visible today, just upriver from St. Louis. Along the edges of the quarry grow trees of the same kind of forest that Clark observed. The woodlands along the Missouri River were cut heavily in the nineteenth century for many reasons, not the least of which was to provide fuel for steamboats and, after 1855 when railroads reached Iowa, wood for railway ties. It took about eight hundred acres— that's more than a square mile—of good oak woodland to provide a mile of railroad ties, and the ties lasted less than ten years. What is now the state of Iowa had about nineteen million acres of forests—little of it, however, along the Missouri; most was in the northeast of that state, back from the Mississippi. It is estimated that by 1900, four million acres of forest had been felled for logging, cattle grazing, and coal mining. On the other hand, farmers planted trees for shelter belts and windbreaks, and trees were planted in cities and towns.

Edible Plants

The Missouri River valley provides people with many benefits, and Lewis and Clark took advantage of many of these as they passed along the limestone bluffs between the modern locations of Jefferson City and Columbia, Missouri. On June 5, York (Clark's slave and the only black member of the expedition) "Swam to the Sand bar to geather greens for our Dinner and returnd with a Suffecent quantity wild *Creases* or Teng grass," Clark wrote. Wild cress remains a common plant along the Missouri floodplain.

They enjoyed the sounds and sights, as travelers along this part of the river do today. Repeatedly they referred to "Butifull" prairies approaching the river and extending back from it. Near Overton, Missouri, Clark wrote that there was "delghtfull land." Near Jefferson, where the expedition camped on June 4, Clark named a small stream Nightingale Creek because of the beautiful sounds of the birds calling all night—probably the whippoorwill, which you can still hear echoing in this region on a spring evening.

While Clark wrote many of the daily notes, Lewis focused on aspects of the environment that seemed of particular interest. In this section of the journey, he became interested in the plants the Indians used as food, especially those that grew in or near the water. He commented on several of these plants, describing not only the plants but the method of preparation and the flavors. He found that these water plants were important foods for the Indians, and he described the use of *Nelumbo lutea*, which we call American lotus. "The Kickapoo calls a certain water plant with a large Circular floating leaf found in the ponds and marshes in the neighbourhood of Kaskaskias & Cahokia—Po-kis'-a-co-mash, of the root of this plant the Indians prepare an agreable dish, the root when taken in it's green state is from 8 to 14 inches in circumpherence." The Indians collected the roots in the fall, walking in the water and feeling with their feet for the roots in the muddy bottom. After washing the roots and scraping off the black "rind," the Indians boiled cuttings of root to make "an agreeable soupe," Lewis wrote. They also dried the root, either in the sun or by "expos[ing] it to the smoke of their chimnies." Lewis wrote that the dried root remains edible for several years. "It is esteemed as nutricius as the pumpkin or squash and is not very dissimilar in taste—The Chipiways or sateaus call this plant

American Lotus, Used by the Indians of Missouri as Food. Its roots and flower were edible, and Lewis describes them as tasty. © *W. D. Bransford. Contributed by Lady Bird Johnson Wildflower Center.*

Wab-bis-sa-pin or Swan-root—The french or Canadians know it by two names the Pois de Shicoriat or Grane de Volais."

Lewis wrote that the flower sets a conelike fruit that was also eaten by the Indians: "The surface of the cone when dryed by the sun and air after being exposed to the frost is purforated with two cicular ranges of globular holes from twenty to 30 in number which forms the center placed at the distance of from an eigth to 1/4 of an inch assunder, each of those cells contains an oval nut of a light brown colour much resmbling a small white oak acorn smothe extreemly heard, and containing a white cernal of an agreeable flavor; these the native frequently eat either in this state or roasted; they frequently eat them also in their succulent state."

This plant, he observed, is also eaten by wildlife. "The bear feed on the leaves of this plant in the spring and summer—in the autumn and winter the Swan, geese, brant, ducks and other aquatic fowls feed on the root."

During this time, Lewis also described the use of the American groundnut, *Apios americana,* which he noted is also called Indian potato and potato-bean. This is a legume and therefore a relative of peas, beans, and peanuts. He noted that it grows along streams. He described its preparation:

[T]his they boil untill the skin leaves the pulp easily which it will do in the course of a few minutes. The outer rind which is of a dark brown coulour is then scaped off the pulp is a white coulour, the peetatoe thus prepared is exposed on a scaffold to the sun or a slow fire untill it is thoroughly dryed, or at other times strung upon throngs of leather or bark and hung in the roofs of their lodges where by the influence of the fire and smoke it becomes throughly dryed, they are then prepared for use, and will keep perfecdtly sound many years, these they boil with meat or pound and make an agreeable bread . . . . This pittaitoee may be used in it's green or undryed state without danger provided it be well roasted or boiled.

The groundnut is one of the most famous edible wild plants of the eastern United States and was a major food of the Indians. It also was important to early European settlers and explorers. Although it is nutritious—containing a greater percentage of protein than the potato—attempts to use it as a crop in European-based agriculture have not succeeded, probably because the plant needs two to three years to mature, a

American Groundnut, Food for Indians and Early European Settlers in North America, considered by some biologists to be the most edible plant in North America. © *George H. Bruso. Contributed by Lady Bird Johnson Wildflower Center.*

cycle inconsistent with methods used with other crops. But attempts to domesticate it continue, so it may be a food for our future, reducing the need for the addition of nitrogen fertilizer, since, like all legumes, ground-nut has symbiotic relationships with bacteria and fungi that fix nitrogen.

Finally, Lewis described a plant "called by the Chipeways Moc-cup-pin"—the water lily (*Nymphaea tuberosa*)—as "another root found in mashey lands or ponds which is much used by the Kickapoos Chipaways and many other nations as an article of food." Common in Lewis's time, this plant is essentially extinct in Missouri today, with a single possible sighting in the last ten years and a last collection made in Missouri around 1897. "In it's unprepared state," he wrote, it "is not only disagreeable to the taste but even dangerous to be taken even in a small quantity; in this state it acts as a powerfull aemnetic. A small quantity will kill a hog yet prepared by the Indians it makes not only an agreeable but a nutricious food."

The preparation involved cutting the root into pieces and burying it with pieces of wood. This rough sort of kiln was then heated so that the wood burned slowly. Afterward, the dried root could keep a long time. Then it was "either boiled into a pulp in their soupe or elss boiled [and eaten] with bears oil or venison and bears flesh," or made into a flour and then into a bread.

As the expedition fought its way upstream against the wild and ever-changing Missouri River, they found that the vegetation along the river was important to them—creating beautiful scenery; providing timber, wildlife habitat, and edible plants. Lewis and Clark were coming to understand the patterns along the river and the forests beyond. They were beginning to understand the processes that created those patterns. This understanding was a gateway to restoring the river, a gateway that we will examine in later chapters.

4

INTO THE TALL-GRASS PRAIRIE

"One of the most butifull Plains, I ever Saw" where "nature appears to have
exerted herself to butify the Senery by the variety of flours" which "profumes the Sensation, and amuses the mind."

—WILLIAM CLARK, JULY 4, 1804,
THE PRAIRIES ALONG THE MISSOURI RIVER AT THE
LOCATION OF MODERN ATCHISON, KANSAS

ON JUNE 26, 1804 the expedition camped just above the mouth of the
Kansas River, at the present location of Kansas City, Missouri. The
men spent several days there because they had to repair the pirogue, which
they emptied, brought up on land, and turned over. During this work, on
June 28, the expedition saw buffalo, which they did not kill, but which indicated that the expedition had entered a second ecological region, the tall-
grass prairie, and left behind the eastern deciduous forest of Missouri.

Clark praised the location for its beauty and its opportunity for defense.
He wrote on June 28 that the Kansas Indians lived "in a open & butifull
plain" and that "the high lands Coms to the river Kanses on the upper Side
at about a mile," which made "a butifull place for a fort, good landing
place."

Just downstream, on June 25, the expedition had camped in what is now
Sugar Creek, a suburb of Kansas City, and saw many deer "feeding on the
young willows & earbage in the Banks and on the Sand bars in the river." It
was a productive location, with "Plumbs Raspberries & vast quantities of
wild apples." The next day the men killed seven deer. There they also saw "a

great number of *Parrot Queets*," the Carolina parakeet that is now extinct, but for which Lewis and Clark provided the first written observation west of the Mississippi. It was a place of appealing biological diversity.

The site was an important urban location for geographic reasons: it is the confluence of a major tributary of the Missouri, always a good city location. At the junction with the Kansas River the Missouri's riverbed changes direction and upstream lies north-south. So, this was the farthest west one could go upstream on the Missouri below Sioux City, above which the riverbed again lies east-west. A traveler's options at the Kansas River mouth were to take that smaller river, which few did; to go north to Omaha and take the Platte River west, the route that later became most popular; to follow Lewis and Clark and continue up the Missouri; or to travel by land. These factors made the location a natural one for a city as well as a fort.

A Vast Ecosystem

At the time of the Lewis and Clark expedition, the prairies extended in one vast sheet west from Indiana to the Rocky Mountains, covering more land area than any other ecosystem in North America. Prairies reached north to Canadian tundra, covering much of Saskatchewan and Alberta; then extended south through eastern Montana, the Dakotas, Minnesota, Nebraska, Iowa, western Illinois and Ohio, eastern Wyoming and Colorado, Kansas, western Missouri, eastern New Mexico, Oklahoma, and Texas; ending finally at the desert's edge in Arizona. Separate prairie outliers extended in the far West: the Palouse grasslands of Washington, which Lewis and Clark would see, and grasslands of California's Great Central Valley, south of where the expedition passed. These vast and often seemingly empty lands were the home of the Apache, Assiniboine, and Cheyenne; the Chippewa, Comanche, and Crow; the Kiowa, Mandan, Omaha, and Osage; and the Otee, Pawnee, Ponca, Sioux, and Wichita.

The prairie landscape impressed Lewis and Clark as it would so many travelers who followed. On August 25 Lewis and Clark and a small group of their men traveled by foot to visit the Spirit Mound, a mound that the Indians had told them was inhabited by "*little people* or Spirits." They hiked six miles from the Missouri River, across its floodplain. Clark measured it, as he tended to do with everything on the landscape, and found it to be about

300 yards by 60 or 70 yards "from the longer Side of the Base it rises from the North & South with a Steep assent to the hight of 65 or 70 feet, leaveing a leavel Plain on the top of 12 feet in width and 90 in length." They were able to get a good view of the tall-grass prairie. Clark wrote that "from the top of this Mound we beheld a most butifull landscape; Numerous herds of buffalow were Seen feeding in various directions, the Plain to North N W & N E extends without interuption as far as Can be Seen—"

Trees were few. "No woods except on the Missouris Points if all the timber which is on the Stone Creek was on 100 a[c]res it would not be thickly timbered," but "the Soil of those Plains are delightfull."

Clark also saw "Great numbers of Birds." His analysis of the legend of the Spirit Mound demonstrates that he, like Lewis, was an excellent naturalist, able to observe, analyze, and understand. Clark wrote that he saw these birds catching "a kind of flying ant which were in great numbers abought the top of this hill." He observed further that "The Surrounding Plains is open void of Timber and leavel to a great extent: hence the wind from whatever quarter it may blow, drives with unusial force over the naked Plains and against this hill." He then considered the implication of

The View Today from the Spirit Mound Visited in 1804 by Lewis and Clark. Neat agriculture fields fill the landscape that was then beautiful prairie. *D. B. Botkin.*

the winds pushing the insects: "the insects of various kinds are thus involuntaryly driven to the mound by the force of the wind, or fly to its Leward Side for Shelter; the Small Birds whoes food they are, Consequently resort in great number to this place." He thought some more about his observations and wrote, "One evidence which the Inds Give for believeing this place to be the residence of Some unusial Spirits is that they frequently discover a large assemblage of Birds about this mound." Aha—he arrives at a possible explanation for why it is known as the Spirit Mound. The winds blow the insects to the lee; the birds learn this and concentrate there; and the Indians, observing the concentration, see this as a supernatural event.

Once again this careful thinking about the landscape takes place in difficult circumstances—a hike six miles inland during the heat of the summer, which caused Lewis to become "much fatigued from heat," according to Clark, who added, "Several of the men complaining of Great thirst, deturmined us to make for the first water which was the Creek in a bend N. E. from the mound about 3 miles." They had to walk through the prairie an additional three miles simply to get a drink of water. Clark's ability to observe, think, analyze, and later write down his thoughts suggests abilities uncommon in our age, when few of us are used to experiencing life within nature the way Lewis and Clark did. We might be, like Lewis and several of the men, desperately thirsty to the point where we could think of little else, but Clark carries on, measuring, reflecting, and writing. It is a kind of heroism we are not accustomed to celebrating, nor even used to imagining.

Within the tall-grass prairie, the easternmost portion of North America's grasslands, grew two hundred plant species, including such major ones as bluestem, dropseed, compass plant, coneflower, and various gentians. The vast prairie impressed many travelers, including the famous nineteenth-century historian Francis Parkman, who wrote in 1847 as he left Fort Leavenworth (near Independence Creek, named by Lewis and Clark because they camped there on July 4, 1804), "The scenery needed no foreign aid. Nature had done enough for it." He particularly liked the "alternation of rich green prairies and groves that stood in clusters," that is, the American savannah. It was an almost domesticated beauty. Along the stream banks, Parkman found "all the softened and polished beauty of a region that has been for centuries under the hand of man."

Another early prairie traveler, Josiah Gregg, wrote in 1831 that "not a single landmark is to be visible for more than forty miles—scarcely a visible eminence by which to direct one's course." Indeed, "All is as level as the sea, and the compass was our surest, as well as principal, guide."

North American prairies produced some of the best soils in the world, in part because of geological and climatic events there led to the deposition of soil particles good for crops, and in part because prairie vegetation produces a deep organic layer as roots, rhizomes, stems, leaves, flowers, and fruits decay. These soils are much better for agriculture than those of the lands to the east, such as those in New England, where much of the soil is rocky because of glacial depositions—hence the old New England

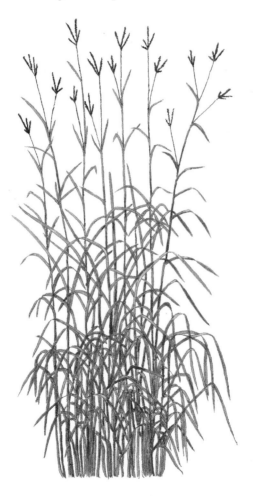

Big Bluestem, One of the Major Prairie Grasses.
Courtesy of Fermi National Accelerator Laboratory.

It Is Hard to Find Original Prairie. So hard that one of the best places to look is an old cemetery, because there the land may never have been plowed and as a result the original prairie vegetation could hold on. This picture, of Rochester Cemetery in Cedar County, Iowa, shows one of these tiny prairie remnants. *Craig Kohl, Photographic Services, University of Iowa.*

saying, "You get two crops of rock and one crop of potatoes every year." As a result, once the Erie Canal opened and travel westward became safer and more convenient, many farmers from New England moved to the prairie states. The migration of farmers from the poor soils of New England to better soils westward led to New Hampshire's population reaching its peak in 1860 and continuing to decline until late in the twentieth century.

Today it is difficult to find *any* prairie along the pathway of Lewis and Clark. Of all the ecological regions, this has been the most altered, done in by that common—but, from the prairie's point of view, exotic—device, the steel plow.

Potential for Agriculture

On July 4, 1804, the expedition camped near the location of modern Atchison, Kansas. Clark went out into the prairie and later wrote in his journal that this was "one of the most butifull Plains, I ever Saw, open & butifully diversified with hills & vallies all presenting themselves to the river covered with grass and a few scattering trees." The creek was "handsom," he wrote, and "The Plains of this countrey are covered with a Leek Green Grass, well calculated for the sweetest and most norushing hay," which suggests that Clark was thinking of uses to which the land might be put—a wonderful potential for farming. The land was already producing much to eat: "Groops of Shrubs covered with the most delicious froot is to be seen in every direction, and nature appears to have exerted herself to butify the Senery by the variety of flours Delicately and highly flavered raised above the Grass, which Strikes & profumes the Sensation, and amuses the mind." The grassland was "interspersed with Cops of trees," and the trees spread "ther lofty branchs over Pools Springs or Brooks of fine water."

The country was so beautiful that Clark wrote that the mind was thrown "into Conjecterng the cause of So magnificent a Senerey in a Country thus Situated far removed from the Sivilised world to be enjoyed by nothing but the Buffalo Elk Deer & Bear in which it abounds." It was one of the few times that Clark departed from his usually direct reporting style and lists of measurements to wax philosophical.

Since it was Independence Day, Lewis and Clark named the small tributary of the Missouri by which they camped "Independence Creek," and they "ussered in the day by a discharge of one shot from our Bow piece"—their cannon.

Today Atchison, Kansas, is heavily developed, and the surrounding countryside has been primarily farmland for a long time. At the mouth of Independence Creek is a narrow but pleasant park and boat ramp. This tributary, like many along the lower Missouri, has been channelized, and levees have been built along its edges to protect the farmland. But the mouth has been straightened so that the creek no longer enters the Missouri River exactly where Lewis and Clark saw it.

On that day, Lewis walked to the top of a "high moun" to take in an "extensive view" that included "great numbers of Goslings." Today one can walk up the bluffs on city streets to the campus of Benedictine College. Along the edge of the bluff of this pleasant campus, probably the same summit where Lewis stood, one can sit on a park bench and look out to Independence Creek, Benedictine Bottoms, and the Missouri River beyond. Benedictine Bottoms contains a large section of row-crop plantings on the bottomland that Clark so admired as a site for farming, and another section that is in an early stage of restoration to wildlife habitat.

Since the time of Lewis and Clark, the Missouri River basin has become one of the nation's major agricultural areas: about 95 percent of the basin's use is for agriculture. The emphasis along the lower Missouri, downstream from Gavin's Point Dam, is on row-cropping; west and upstream, past the hundredth longitudinal meridian, the predominant use shifts to grazing. This large-scale agriculture has been a great benefit to the United States, but it also has brought changes of an invisible kind unimaginable by Lewis and Clark and their contemporaries: the introduction of artificial chemicals used as pesticides and the increase in levels of nitrate and phosphate from widespread application of fertilizers.

Agriculture and Water Quality

Before the channelization of the Missouri River, measurements of water quality were few and scattered, but they provide us with some baseline information, suggesting that by 1984 nitrate and phosphate concentrations in the Missouri River below the big dams had increased to four times the baseline level. Meanwhile, upstream in the reservoirs, nitrate and phosphate concentrations seem to have decreased.

This downside of agriculture became apparent in 1964, when a fish kill extended more than a hundred miles downstream from Kansas City, Missouri. Monitoring of pesticides remained spotty, but between 1968 and 1976, fish flesh at Council Bluffs, Iowa, was found to have concentrations of the pesticide dieldrin that exceeded public health standards in 13 percent of the samples, and DDT and its breakdown products exceeding public health standards in one-third of the samples. In the early 1970s, PCBs, aldrin, and dieldrin in fish analyzed at Hermann, Missouri, posed a potential health threat. In the mid-1970s, dieldrin levels were high enough in catfish that the Missouri Department of Conservation issued warnings. People stopped buying catfish, affecting commercial fisheries.

Pesticides entered the river not just from agriculture but from urban, suburban, and industrial use. By 1984, chlordane, at the time commonly used to kill household pests such as termites, exceeded established safe levels in fish in the lower Missouri. In 1987, the Missouri Department of Health advised against consumption of specific commercial fish species from certain areas of the Missouri River because of contamination with toxic compounds.

About seven hundred million pounds of pesticides, representing more than one hundred compounds, are applied nationwide, and herbicides account for about 60 percent of the total pesticides found in the nation's waters. Public health standards and environmental-effects standards have been established for some but not all of these compounds. And there is no major monitoring program for the pesticides flowing down the main channel of the Missouri River.

Elsewhere, monitoring the levels of these chemicals in the waters has increased greatly. The United States Geological Survey has established a network for monitoring sixty watersheds throughout the nation. One of these is the Platte River, one of the major tributaries of the Missouri. The most common herbicides used for growing corn, sorghum, and soybeans along the Platte River were alachlor, atrazine, cyanazine, and metolachlor, all organonitrogen herbicides. Monitoring on the Platte near Lincoln, Nebraska, has suggested that during heavy spring runoff, concentrations of some herbicides may be reaching or exceeding established public health standards. But this research is just beginning, and it is difficult to draw definitive conclusions about whether present concentrations are causing harm to public water supplies or to wildlife, fish, fresh-water algae, or vegetation. The advances in knowledge tell us more and more regularly, about how much of many artificial compounds are in the waters, but their environmental effects are still unclear.

Here is a potentially major national problem that has received relatively little attention: most of the focus on pollution has been on urban and industrial areas and on feedlots, with much less on pollution from row-crop agriculture. We are conducting experiments with nature without the usual qualities of the scientific method: treatments and controls, adequate monitoring, and long-term experimentation in laboratories.

The view from the bluff at Atchison is a mixed one. The scene below is a combination of farmlands, wetlands, streams, rivers, and heavy land development. It is no longer a scene to rhapsodize about, the way Lewis and Clark did in 1804.

The problem is not that the land has been converted to benefit people. It is that these changes seem to have been made in large part without the kind of care for design and beauty that one would expect to take in the garden-like surroundings that Clark found.

Nature and Prairies

Fires and grazing maintain prairies. Tall-grass prairie, which grows in the easternmost, wetter prairie region, is replaced by trees and shrubs when there are no fires and no grazing. In the westernmost prairie region, short-grass prairie gives way to desert shrubs if there are no fires or grazing. Changes—alterations to the current condition—are necessary for the persistence of the prairie and its hundreds of species of grasses and nongrassy flowering plants, commonly called "forbs."

Wetlands were part of the prairie, especially to the north of the Missouri in land that had been glaciated. Sometimes large, sometimes merely small "potholes," these wetlands were drained by farmers. They had been nesting and feeding areas for many birds, whooping cranes, trumpeter swans, ducks, geese, and blackbirds among them. Muskrats shared these habitats with turtles, fish, frogs, and salamanders.

In the partially settled countryside of Missouri and Iowa, wildlife declined after the time of Lewis and Clark. Many small streams in Iowa were straightened, channelized, and shortened, making them unsuitable habitat for the otters and beavers that had once thrived there. Heavily trapped in the seventeenth century, these animals were in decline in Iowa by the time of Lewis and Clark.

Change came rapidly to what is now Iowa, beginning less than twenty years after Lewis and Clark returned through this countryside. European settlement began in 1833, with the Black Hawk land purchase, six million acres to the west of the Mississippi River. Many of the larger mammals disappeared from this state in the nineteenth century—mountain lions eliminated by 1867; elk by 1871; bears by 1876; and wolves about a decade later. Passenger pigeons had been pretty much eliminated from Iowa by the 1890s and would become extinct by 1914.

Iowa became a state in 1846, forty years after Lewis and Clark traveled through it, and when there were about a hundred thousand residents. Today, the state is home to almost three million people, and more than 95

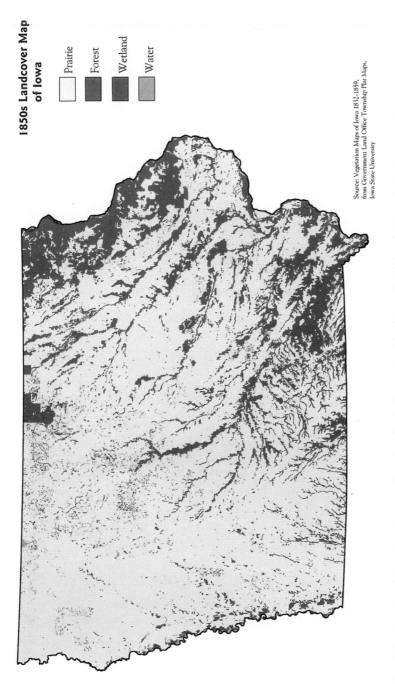

1850s Landcover Map of Iowa

- Prairie
- Forest
- Wetland
- Water

Source: Vegetation Maps of Iowa 1832-1859, from Government Land Office Township Plat Maps, Iowa State University

Map of the Extent of Prairie and Forest in Iowa About 1850. The land is mostly prairie, with woodlands along the Missouri River (the west edge of Iowa), along the major rivers, and in the northeast. *From* IOWA—Portrait of the Land, *Iowa Dept. of Natural Resources, Iowa, 2000. p. 20.*

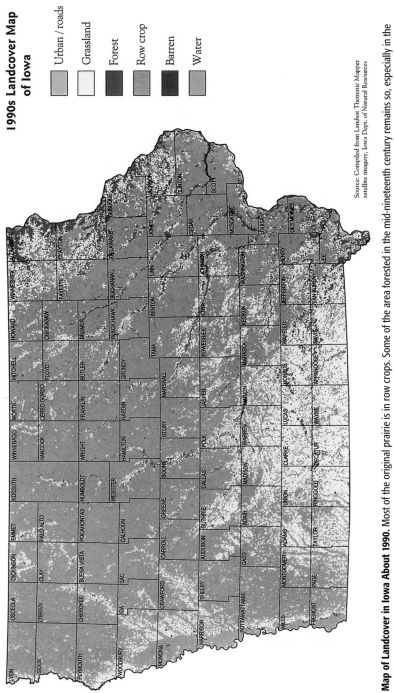

1990s Landcover Map of Iowa

Urban / roads
Grassland
Forest
Row crop
Barren
Water

Source: Compiled from Landsat Thematic Mapper satellite imagery, Iowa Dept. of Natural Resources

Map of Landcover in Iowa About 1990. Most of the original prairie is in row crops. Some of the area forested in the mid-nineteenth century remains so, especially in the northeast. Grasslands are primarily pastures, not native prairies. *From IOWA—Portrait of the Land, Iowa Dept. of Natural Resources, Iowa, 2000. p. 21.*

percent of its prairie and two-thirds of its woodlands are in agriculture, according to *IOWA: Portrait of the Land: A Century of Change*. The major changes in vegetation of Iowa occurred in less than one human lifetime, about sixty or seventy years.

The Konza Prairie

One has to search far to find prairie and in general must go quite a distance from the Lewis and Clark trail to find any. But it can be found just outside of Manhattan, Kansas, at the Konza Prairie, one of the largest remaining prairies in the United States, covering 8,616 acres. Prairie is so rare that the preserve is owned by the Nature Conservancy and Kansas State University and used as a research station, part of a National Science Foundation program for long-term ecological research. Like the vast prairie lands that Lewis and Clark saw, Konza contains more than grasslands—it also features forests, shrublands, and streamside vegetation.

Konza Prairie Today. *D. B. Botkin.*

Beaver and the Fur Trade

Beaver were an important part of the North American fur trade long before
the Lewis and Clark expedition. Jefferson had a scientific curiosity about
wildlife, plants, and geology, but he was also interested in the commercial
potential of natural resources. Two questions Lewis and Clark sought to
answer were whether beaver were abundant along their route, and if so,
whether the United States might begin to take over from Great Britain
some of the beaver trade with the Indians.

Despite their interest in this species, Lewis and Clark did not record any
observations of beaver until July 3, 1804, when they reached the neighbor-
hood of Leavenworth and Atchison, Kansas. On that day Clark recorded
that they stopped at a deserted old trading house where they "found a verry
fat horse, which appears to have been lost a long time," and passed a large
island called "*Isle la de Vache* or *Cow Island*." On the shore was "a large Pond
Containg Beever." That trading house marked the first observation of
beaver and suggests that Lewis and Clark were still within countryside
known and used to some extent by trappers, who had exploited and pretty
much eliminated beaver downstream. Like other wild living resources,
beaver were generally perceived at the time as things to be exploited but not
conserved, harvested but not sustained, and they were disappearing with
the inroads of European-based settlement.

Lewis and Clark next saw beaver near Council Bluffs, just north of modern Omaha. Afterward, they saw these animals frequently. They caught a
few in the fall when they had reached the Mandan villages where they
would spend the winter.

The next spring, on April 12, 1805, just after the expedition had left the
winter camp near the Mandan villages, and when Lewis and Clark were not
far west of modern Bismarck, North Dakota, and near the mouth of the
Little Missouri River, Lewis wrote that "George Drewyer shot a Beaver this
morning, which we found swiming in the river." Considering this, he added
"the beaver being seen in the day, is a proof that they have been but little
hunted, as they always keep themselves closly concealed during the day
where they are so." Then on April 13, 1805, Clark wrote that they "Cought 3
beaver this morrning." And on April 16, Lewis wrote "there was a remarkable large beaver caught by one of the party last night. These anamals are
now very abundant." Lewis and Clark's observations are so good that they

tell us that beaver were once plentiful on streams where they have been gone for so long that even their effects on the vegetation are no longer apparent, and it is sometimes hard to believe beaver were ever there.

Today much of the habitat along the Missouri that beaver might have used is gone—the backwaters and the bottomland forests. Even many of the tributaries have been channelized and have levees along them. When beaver do return, they are considered pests whose dams flood land that people want for other uses. So to see beaver on a Lewis and Clark journey, you will have to travel farther upstream to less intensively developed areas.

Buffalo on the Prairie

In addition to writing descriptions of new species, Lewis and Clark often wrote about the abundance of wildlife. When they visited the Spirit Mound, Clark noted that he saw upwards of eight hundred buffalo and elk feeding.

The expedition saw its first bison on June 6, 1804, when the men were just west of Columbia, Missouri (somewhat south of where Interstate 70 crosses the Missouri River today). Bison were sighted again on June 28, when the expedition reached the Kansas River (at the present location of Kansas City). Once across the Kansas River, Lewis and Clark began to see herds of bison moving across the plains. By now the expedition had passed the settled countryside of the Missouri and had journeyed into the plains and into buffalo country.

The men of the expedition did not shoot a bison until August 23, more than two months after the first sighting. By that time, they were in South Dakota, about four hundred miles upriver from their first encounter (near the present towns of Yankton and Vermillion). It was not an easy day to go wildlife watching. Clark wrote on that day, "The Wind blew hard West and raised the Sands off the bar in Such Clouds that we Could Scercely See." It was what we would call a sandstorm. He added that "this Sand being fine and verry light Stuck to every thing it touched, and in the Plain for a half a mile the distance I was out every Spire of Grass was covered with the Sand or Dust."

Often the landscape was populated by several species of large grazing animals, spread out before the expedition. On August 22, the expedition was on the Missouri near the Vermillion River—not far from the Spirit Mound, and near the modern boundary between northeastern Nebraska and southeastern South Dakota, where its path is north-south. Clark went on land onto a bluff looking over the river and the surrounding prairie and met two of the men who had been sent out on horseback to hunt. They had two deer. The next day, Clark "went out and Killed a fine Buck, J. Field Killed a Buffalow," and in addition "2 Elk Swam by the boat whilst I was out." He saw several coyotes (which he called "Prarie wolves"). Lewis meanwhile "Killed a Goose." Game was everywhere—on the land and in the water.

For the rest of the summer and fall, they were in buffalo country and often found these animals in abundance. When they reached the Big Bend of the Missouri—now the site of South Dakota's Big Bend Dam, a location famous long after their journey—Clark went for a walk on September 20 to examine the bend from high ground, going up "high irregular hills." The next day, he noted that there were "delightfull plains with graduel assents from the river" where he saw "Great number of Buffalow Elk & Goats feed." Once again, it was not an especially good day for nature-watching, as during that night Clark was wakened by noise and "by the light of the moon observed that the Sand was giving away both above & beloy and would Swallow our Perogues in a few minits." He had the men push the boats offshore into the main channel, adding that "we had not got to the opposit Shore before pt. of our Camp fel into the river." In typical Lewis and Clark fashion, natural history observations are made despite hardships, from the river's banks collapsing to—later in their journey—intense storms, some with hail, some lasting days; temperatures well below freezing; and limited food supply.

And so again, on September 22, Clark noted that "a Thick fog this morning detained us untill 7 oClock." But he wrote that once the fog lifted, "The plains on both Sides of the River is butifull," and there were "noumerous herds of Buffalow to be Seen in every derections." American buffalo were in great abundance in the presettlement landscape.

As they moved northward into what is now North Dakota, the bison continued to appear on the plains in great numbers. On October 19, when the expedition was near to the Mandan villages where they would spend the winter, Clark "walked out on the Hills" along the river and "observed Great numbers of Buffalow feedeing on both Sides of the river." Rarely did

Lewis and Clark try to count the large populations of buffalo they saw. On the return trip, however, Lewis did attempt an estimate. The expedition had just recently separated into three groups: one with Lewis to explore the Marias River; a second with Clark to explore the Yellowstone; and a third to return down the Missouri. The three groups would meet farther downstream. On July 11, 1806, Lewis was on the Missouri River near White Bear Island, where he had been the previous July 16. He wrote that "the morning was fair and the plains looked beatifull," and added that "the air was pleasant and a vast assemblage of little birds which croud to the groves on the river sung most enchantingly." It was the rutting season for the buffalo, and the bulls kept up "a tremendious roaring we could hear them for many miles and there are such numbers of them that there is one continual roar." He continued, "I sincerely belief that there were not less than 10 thousand buffaloe within a circle of 2 miles."

Buffalo continued to be abundant on the Great Plains for decades after Lewis and Clark's expedition. All the early writers tell of immense herds of bison, but rarely do they count them. An exception is General Isaac I. Stevens, who on July 10, 1853, was surveying for the transcontinental railway in North Dakota. He and his men climbed a high hill and saw "for a great distance ahead every square mile" having "a herd of buffalo upon it." He wrote that "their number was variously estimated by the members of the party—some as high as half a million. I do not think it any exaggeration to set it down at 200,000."

One of the better attempts to estimate the number of buffalo in a herd was made by Colonel R. I. Dodge, who took a wagon from Fort Zarah to Fort Larded on the Arkansas River in May 1871, a distance of thirty-four miles. For at least twenty-five of those miles, he found himself in a "dark blanket" of buffalo. Dodge estimated that there were 480,000 in the mass of animals he saw in one day. At one point, he and his men traveled to the top of a hill, where he estimated that he could see six to ten miles, and from that high point there appeared to be a single solid mass of buffalo extending over twenty-five miles. At ten animals per acre, not a particularly high density, there would have been between two and a half to eight million animals. These estimates suggest that Lewis was probably not far off in his guess that there were 10,000 within a two-mile circle.

Just before and just after the Civil War, "there were always buffalo somewhere" along the tracks of the Kansas Pacific Railroad. In the fall of 1868, "a train traveled one hundred twenty miles between Ellsworth, Kansas and Sheridan, Wyoming through a continuous, browsing herd, packed so thick that the engineer had to stop several times, mostly because the buffaloes would scarcely get off the tracks for the whistle and the belching smoke." That spring a train had been delayed for eight hours while a single herd passed "in one steady, unending stream." We can use such reports to set bounds on the possible number of animals seen. At the largest extreme, we can assume that the train bisected a circular herd with a diameter of 120 miles. Such a herd would cover 11,310 square miles, or more than seven million acres. Suppose people exaggerated the density of the buffalo, and there were only ten per acre—a moderate density of a herd. This single herd would have numbered seventy million animals!

The demise of the buffalo is famous. We think of buffalo as creatures of the American West, but they were found over a much wider range at the time of early European discovery and exploration of North America. By the time of the Lewis and Clark expedition, the geographic range of the buffalo had been greatly reduced—something few Americans know.

Lewis and Clark's first sighting took place once they were past the European settlement, where buffalo had once roamed but been eliminated. This was the characteristic pattern: buffalo were hunted to local extinction or driven out as farms with fences and grazing lands for European cattle were established. The plow and the buffalo were considered incompatible.

It is little known how widely dispersed buffalo were at the time of European discovery and early settlement of North America. Cabeza de Vaca, the famous Spanish castaway who spent eight years with the Indians in the 1530s and later recounted his experiences when he returned to Spain, saw buffalo in southern Texas in the 1530s. In 1612 explorer Samuel Argoll sailed on the Chesapeake Bay and saw "a great store of cattle" that were "heavy, slow, and not so wild as other beasts in the wilderness." He had seen buffalo. In 1701 there was an attempt to domesticate buffalo in a new settlement on the James River in Virginia. Near Roanoke, Virginia, buffalo were common at a salt lick until the mid-eighteenth century. One herd was reported in southwestern Georgia in 1686. Buffalo were killed off in Georgia by 1780, in South Carolina by 1775. Some evidence suggests that the buffalo had only recently reached the East Coast when Europeans began to settle and explore that area.

These records and others suggest that before the arrival of Europeans, buffalo may have occupied one-third of North America, reaching as far north as the boreal forests of Canada and as far south as the chaparral of southwestern Texas. Buffalo were found in Canada as far north as Great Slave Lake in the Northwest Territories, to latitude 60 degrees north, just north of the present Wood Buffalo National Park, which lies in northern Alberta, and the boundary of the Northwest Territories. Fossils of bison, some as old as 40,000 BC, have been found from New Jersey to California.

The destruction of the buffalo took place with a rapidity that is hard to grasp. They were killed for two reasons: for profit and to eliminate the food of the Indians—ecological warfare. Colonel R. I. Dodge, the same person who later made one of the estimates of the numbers in a herd, was quoted in 1867 as saying, "Kill every buffalo you can. Every buffalo dead is an Indian gone." In 1864 buffalo robes and tanned hides began to be shipped from St. Louis eastward. New technologies made it possible to increase the use of buffalo. Modern rifles made it easy to kill them. Trains made transport of hunters and hides easier. A new tanning process, developed in Germany, allowed the treatment of many more hides, and the finished hides provided a better, more desirable grade of leather. A European market opened for the improved hides.

In the first half of the nineteenth century, the animals were seen as a commodity, like gold, to be removed as quickly as possible for individual profit. Railway construction crews spent their winters, when it was not possible to work on the railroads, hunting buffalo. Ironically, although many saw buffalo as a way to riches, few of the buffalo hunters got rich.

The Civil War had its effect on the buffalo. During the war, buffalo hides were used by the military, increasing the market. After the war, army veterans, skilled in shooting rifles, headed west, where they used these skills against the buffalo. A major increase in exploitation of many of America's biological resources occurred just after the Civil War, as new lands opened up in the West, as displaced Southerners found their way westward, and as our nation shifted away from war to the development of transcontinental railways and the settlement of the West. Wild Bill Hickock became one of the major buffalo hunters, along with Buffalo Bill Cody.

Records of the number of buffalo killed were neither organized nor well kept, but enough are available to give us some idea of the number taken. The Indians were also killing large numbers of buffalo for their own use and for trade. Estimates range up to 3.5 million buffalo killed each year during the 1850s. In 1870 about two million buffalo were killed. In 1872 one company in Dodge City handled 200,000 hides. Estimates based on the sum of reports from such companies and guesses at how many more would have been taken by small operators and not reported suggest that about 1.5 million hides were shipped in 1872, and the same number in 1873. In these years, buffalo hunting was the main economic activity in Kansas.

As this high harvest continued, concern about the possible extinction of buffalo grew. In 1871 the U.S. Biological Survey sent George Grinnell to survey the herds along the Platte River. He estimated that there were only 500,000 buffalo there and that, at the present rate of killing, the animals would not last long. As late as the spring of 1883, a herd of an estimated 75,000 crossed the Yellowstone River near Miles City, Montana, but it was estimated that fewer than 5,000 reached the border. By the end of that year—only fifteen years after the Kansas Pacific train was delayed for eight hours by a huge herd of buffalo—only a thousand or so buffalo could be found, 256 in captivity and about 835 roaming the plains. A short time later, there were only fifty buffalo wild on the plains. The great era of the plains buffalo was over.

The incredibly rapid demise of these animals demonstrates the power of nineteenth-century technology when put to a destructive purpose. But societal attitudes change. With the rise of the environmental movement in the 1960s, concern with endangered species increased. In recent decades, a revival of interest in buffalo on the plains has begun. Today several national wildlife refuges, parks, and grasslands maintain buffalo, including Fort Niobrara National Wildlife Refuge near to where the expedition passed; the Little Missouri Grasslands; Theodore Roosevelt National Park; and the National Bison Range north of Missoula, Montana. Fort Belknap Indian Reservation of the Gros Ventre and Assiniboine tribes conducts buffalo tours.

And people are beginning to see profit in running buffalo—an idea not discussed by Lewis and Clark, although they—Clark especially—commented frequently about what we call the economic potential of the land and its resources. One of these families, the Masons, who live near Dixon, Nebraska, not far south of where Lewis and Clark shot their first buffalo, started ranching buffalo in 1993. The Masons are returning the land to prairie—in one area, they have planted some prairie grasses and forbs; on

Buffalo Herd on a Private Nebraska Ranch. *D. B. Botkin.*

most of the land, they have let the grass go; and as the buffalo graze, their grazing favors prairie grasses, which are reinvading.

No greater change has taken place in the land west of the Mississippi between the time of Lewis and Clark and today than the demise of the great American prairie and its wildlife. This is less appreciated and less controversial than the cutting of old-growth forests in the Pacific Northwest (where Lewis and Clark would spend their second winter), but its effects are profound. That ocean of grass that was so beautiful to so many travelers is now the world's greatest agricultural land. One need not be judgmental simply to point out the change. But many think and feel that the benefits from agriculture could have come at less cost to the Great Plains and its wildlife, and they believe a bright future would combine both these worlds. Taking this approach would require the clarity of thought, the objectivity, and the toughness, both physical and mental, of Lewis, Clark, and their crew.

5

RESTORING THE LOWER MISSOURI RIVER

O N AUGUST 8, 1804, near present-day Onawa, Iowa, forty miles south of Sioux City, the expedition saw a strange sight. Lewis wrote, "I saw a great number of feathers floating down the river," covering sixty or seventy yards of the river's width. He continued that, "for three miles after I saw those features continuing to run in that manner, we did not percieve from whence they came." It was as if the river had painted itself white. Then he added that "at length we were surprised by the appearance of a flock of Pillican at rest on a large sand bar." Almost three months into the journey, you might think that Lewis was becoming used to strange and extraordinary sights, but miles of white feathers amazed him. There were so many birds that he did not even try to count them, merely writing that the numbers "if estimated" would "appear almost in credible."

He shot one bird as a specimen and made an accurate written description of its features, as Jefferson had instructed him to do. The beak was "a whiteish yellow," and the pouch under the beak was so big that they filled it with five gallons of water. The bird had yellow feet and mostly white feathers except that the "large f[e]athers of the wings are of a deep black." Lewis recognized this as the same pelican that is found in Florida and the Gulf of

White Pelicans Soar Above the Modern Missouri River Near Where Lewis and Clark Saw Thousands.
The white pelican, like its close relative, the brown pelican, suffered greatly from thinning eggshells caused by DDT but have recovered and are once more abundant. *D. B. Botkin.*

Mexico—the white pelican. The birds were "no doubt engaged in procuring their ordinary food," Lewis wrote, "*which is fish.*"

The pelicans were all the more remarkable because, Lewis noted, "we had seen but a few aquatic fouls of any kind on the river since we commenced our journey." He listed a few geese, wood ducks "common to every part of this country," and cranes. The expedition had left too late to see the spring migration of waterfowl, which occurs generally in March or April, and saw primarily resident waterbirds.

Pelicans are just one of many waterbirds whose migration takes them along part of the Missouri River and who depend on the river's habitats. Members of the expedition saw waterbirds on many occasions. "Drewyer killed 2 Deer, Saw great numbers of young gees" is a typical entry, made by Clark on July 19, when they were not far from the location of modern Nebraska City, Nebraska, and therefore nearing Omaha.

The white pelican population declined in the twentieth century because of habitat destruction and because DDT thinned its eggshells, as this chemical did for many birds. Now this species is recovering, but one no longer sees three miles of white pelican feathers on the Missouri River. The change in the status of this species is symbolic of the history of our effects on nature since the time of Lewis and Clark. American society has gone through five stages in its dominant ideas about nature, its use, and the relationship between people and nature. The first, the period of discovery, began with the arrival of Europeans. Lewis and Clark were part of this time of exploration and discovery. During this period, the geographic ranges of many species were reduced, but the impact on entire species, although significant in certain cases, was small in comparison to the events that followed.

The second period was a time of intense exploitation of the continent's wild living resources. Its timing varied somewhat depending on the species, the resources, and the geography, but in general it lasted from the early 1800s to the beginning of World War I. During this period, resources were extracted without much thought about their ability to persist, or if there was some thought about it, little care or attention.

As we will see later, John Jacob Astor, the great fur trader, founded Fort Astoria at the mouth of the Columbia River in Oregon in 1811, creating the first permanent settlement along the northern Pacific coast of what is now the United States, and initiating the exploitation of fur-bearing animals of the Pacific Northwest, which led to the near-demise of the sea otter and the elephant seal.

The demise of the bison took place during this period, as did the conversion of much of the prairie to farmland. East of the Rocky Mountains, large areas of forest were cut—this would happen in Missouri but was especially intense to the north of Lewis and Clark's path, in the Great Lake states where forest resources were more valuable.

Salmon, which Lewis and Clark would find in abundance in the Columbia River system, as we shall also see, were subjected to massive harvesting that peaked in the 1920s.

Professions of formal management for forests and fisheries lay ahead at the beginning of the twentieth century. These were part of the third period, the period of awakening conservation, which began between the turn of the twentieth century and the end of World War I and lasted until the 1960s. Exploitation continued but was accompanied by the first attempts at professional management of forests, fisheries, and wildlife, and the beginnings of a national conservation movement—a political and social movement that had its intellectual antecedents in the nineteenth century. During this period, wilderness management had a single purpose—to maximize the harvest of resources, as if wild, living resources were farm crops or could be treated as such. The United States established its Forest Service at this time, and Yale founded the first forestry school, to train professionals for the new job of maximizing the economic growth of trees.

A fourth period, public environmentalism, began in the 1960s. Public awareness of environmental degradation and the decline of wild, living resources grew, spread, and became established as a social and political movement during this period. Environmentalism developed rapidly, accompanied by the passage of landmark legislation to protect and conserve the environment, such as the Clean Water Act of 1970, the Marine Mammal Protection Act of 1972, and the Endangered Species Act of 1973.

We are now in the beginning of a fifth stage, in which attempts are underway to bring people and nature together once again and to find ways to maintain livelihoods and sustain nature. Some of the initial attempts are taking place today along the route of Lewis and Clark, and, in our attempt to consider nature then and nature now, we have to look also at dominant cultural attitudes and see the cultural forces western civilization applied to nature. Our effects on and relationship to nature are part of the story of the American West. And so it is intriguing to look where Lewis and Clark passed along the Missouri to see what restoration efforts are underway, and whether they are succeeding. Going deeper, we must ask what the goals are, and whether these goals are consistent with Jefferson's European ideas of nature or with the nature that confronted Lewis and Clark as they traveled into the American West. So for a while, we will retrace some of Lewis and Clark's steps and take a look at Missouri River restoration today.

Different species require different kinds of habitat along the river, but most of these habitats were created by the wild, meandering, complex Missouri River, and therefore many have been lost because of channelization, construction of levees and dams, and elimination of much of the seasonal variation in the river's flow. Here, in the great prairies, with rainfall and snowfall decreasing as we move westward, rivers are more and more important to waterfowl as well as to all wildlife and to many plants—especially trees. As I mentioned before, the eastern prairie-forest boundary occurs where rain and snowfall decrease to approximately twenty inches a year—the value varying with latitude, as the cooler the climate, the lower the evaporation of water from the soil. Trees can therefore survive at lower rainfall levels in northern North Dakota than in Nebraska, other things being equal. To the west, when the average rainfall declines to about twelve inches a year, prairie plants can no longer survive, and desert plants take over the landscape. Once again, the average rainfall that leads to this transition changes with latitude.

Many kinds of waterfowl depend on the Missouri River and its backwaters for nesting, breeding, rearing, and migratory feeding. Today, eighteen species of ducks use the river. Of these, wood ducks remain the most common nesting species, just as Lewis reported in his time. There are three geese: Canada, snow, and white-fronted; ten species of wading birds; twenty-five species of shorebirds. A total of two hundred species of migratory birds use the floodplains.

Among species of special concern are the least tern, an endangered species, and the piping plover, a threatened species. Both nest on river sandbars on the Missouri that must be above water level during nesting season. Lewis wrote a long and detailed description of the least tern, two of which he obtained on August 5. "I have frequently observed an acquatic bird in the cours of asscending this river but have never been able to procure one before today," he wrote, suggesting that the bird was relatively common at the time. Observing the birds' habits, he reported that "they lay their eggs on the sand bars without shelter or nest, and produce their young from the 15th to the last of June," a passage that suggests that his observations were careful, consistent, and, in our terms, professional.

The Least Tern, described in detail by Lewis and common at his time, is now an endangered species. © *James M. Zingo.*

He saw that the bird feeds "on small fish, worms and bugs which it takes on the virge of the water." It is a ground bird, he noted, adding that "it is seldom seen to light on trees, an qu[i]te as seldom do they lite in the water."

He described the tern's color and shape, and made measurements: "the weight of the male bird is one ounce and a half, it[s] length from b[e]ak to toe 71/2 inches," he wrote, and continued to list the basic measurements people still make for birds. He noted that the tail had "eleven feathers." His description of the tern's color is elaborate. So in the large and small, Lewis and Clark continually observed, measured, and recorded, even about this tiny bird, now rare.

The attempt to control the Missouri, to treat it as if it were a mechanical plumbing system, has eliminated much of the habitat of water- and shore-birds such as the least tern, other wildlife, fish, and various kinds of vegetation. And as I have discussed, there are now relatively few places a person can go to watch nature on the Missouri River. But recently a number of projects have emerged to recreate these habitats.

Grand Pass Conservation Area

One of the best places to see waterfowl during their migration along the Missouri River is at Grand Pass Conservation Area, five thousand acres managed by the Missouri Department of Conservation, near Arrow Rock, where Lewis wrote about the regeneration of willows and cottonwoods on the sandbars and shores of the Missouri River. Pelicans pass through and stop here, just as they do along other parts of the Missouri where habitat is suitable.

Grand Pass is one of many bottomland areas being restored for fish and wildlife habitat between Sioux City and St. Louis. The website for Grand Pass makes clear its purpose: "Grand Pass is recognized as a waterfowl area, but expect to see many migratory wading birds and shorebirds, in addition to geese and ducks. Habitats on the area include forest, wetland, river island and field. In the bottomland forest, you'll see migratory songbirds and nesting wood ducks. Canada and snow geese come to the wetlands in winter, and shorebirds feed along the river islands and mudflats in spring. Levees, roads and trails at Grand Pass are excellent for bird viewing, especially in February and March." The challenge at each of them is to find the best

Black-crowned Night Heron. © *George Jameson.*

and most economical way to re-create the muted mosaics of backwaters, oxbows, perennial wetlands, seasonal wetlands, wet prairies, and floodplain forests.

One approach to restoring land along the Missouri River is to do as much as possible to speed up natural processes—to take strong measures quickly. The Missouri Department of Conservation is taking this course at Grand Pass Conservation Area. The department has chosen an active, intensive management approach at Grand Pass, creating many artificial wetlands and carefully timing when these are filled and emptied to try to match the natural, pre-channelization seasonal patterns. Pumps move water from the Missouri River to create a variety of seasonal wetland habitats. The pumps are capable of pumping 250 acre-feet a day—enough water to cover 250 acres a foot deep and enough to provide a day's water for more than eight hundred thousand people, more than twice the number of people in St. Louis, at the liberal but average U.S. water use rate of one hundred gallons a day. Water pumped into a wetland is not lost to public use: rather, it is enhanced and then returned to the river.

The water is pumped into large ponds and wetlands separated by levees of different heights, forming an artificial network. About one-third of the area's five thousand acres is in a status referred to as "refuge," meaning once a pond or wetland is constructed there is a "no-touch" policy: no other actions are taken. Refuge areas of no disturbance provide habitat to migrating waterfowl. The refuge wetlands stand in juxtaposition to actively managed areas where much more is done. Some are flooded only in the spring and the fall and then pumped dry in summer and winter, to mimic seasonal wetlands as they used to be on the river. Others are flooded for shorter periods. Still others are restored as wet prairie and planted with switchgrass to protect levees.

The intense management at Grand Pass seems to be working. In recent years, more than 150,000 ducks and more than 50,000 snow geese have stopped at Grand Pass in the spring.

Big Muddy National Fish and Wildlife Refuge

The U.S. Fish and Wildlife Service's major ecological restoration project between Kansas City and St. Louis is the Big Muddy National Fish and Wildlife Refuge. It introduces a new idea about a conservation area. The national parks—an American idea and invention—or nature preserves

Grand Pass Conservation Area After Channelization and Before Restoration of Habitats. *D. B. Botkin.*

have almost always occupied a single, contiguous piece of land and been thought of that way. In contrast, the Big Muddy Refuge is a series of flood plain units spread across the lower Missouri like beads on a necklace of river.

At present, the Big Muddy National Wildlife Refuge consists of seven identified areas that occupy 10,000 acres. The plan is to purchase more land from willing sellers and eventually to establish twenty-five to thirty areas covering 60,000 acres. From Arrow Rock State Park which we visted earlier, one can see two that are already established: Jameson Island and Lisbon Bottom.

Some bends or units of the Big Muddy Refuge will have restored or newly created side channels, called chutes, cut through them to allow some of the river's water to flow away from the main channel and to form quieter side-channel backwaters, wetlands, and other elements of habitat traditionally found in this area.

The idea for such a refuge had been discussed for many years, but the great flood of 1993 revived interest in it, in part because after the flood there

Grand Pass Conservation Area Meander After Restoration of Habitats. A mosaic of habitats, ponds, chutes, uplands, and wetlands, of different shapes and sizes fill the land back from the trees along the Missouri River's shore. *D. B. Botkin.*

were some willing sellers among the landowners—those who had suffered from the flood where the Missouri deposited coarse, sandy soil, impossible for the farmers to use, on many acres of land. The primary purposes of the refuge are to restore natural river floodplain structure and function and to allow better management and conservation of fish and wildlife habitats, and to allow public use, where it is compatible with the primary goals.

Is it necessary, useful, or in the greater public interest to reestablish complex habitats everywhere they once existed? Or can we accomplish what is valuable to nature and people on less than all of the landscape? Along the Missouri River, most desirable refuge lands contain some remnant of floodplain habitats and/or are subject to periodic flooding or drainage problems and thereby are also less valuable as agricultural land. It seems that only portions of the land needed to be returned to the original complex and dynamic habitats—to provide enough spawning, nesting, feeding, and migratory habitat for a large abundance of fish and wildlife.

The light touch of the Fish and Wildlife Service at the Big Muddy—a do-as-little-as-possible-and-let-nature-heal-itself approach—is an alternative to the intensive, active management at Grand Pass. At Big Muddy, the idea is that a few smartly selected and well-executed actions could allow the river to repaint the landscape in the most natural way—a break in a levee at just the right location, for example, could form a single new chute. Then the river would be left alone to erode a complex maze of channels the way it always did and, given the opportunity and following the laws of physics, always will.

The policies at Grand Pass and the Big Muddy are two different approaches to designing landscapes. It is as if two different landscape painters were set before a large canvas. We tend to view environmental issues as a matter of a single truth whose identification is our goal and the solution. Lacking precise information and having only poor understanding of how nature worked in the past, we have no "silver bullets," and it is wise to let a number of approaches bloom on the river.

Fish and Food

On July 24, 1804, Clark wrote that "one of the men cought a *white Catfish*, the eyes Small, & Tale resembling that of a Dolfin." This was probably the channel catfish, and Clark's observations provided the first written description of the species, a species new to western science. The expedition was camped on the eastern, Iowa side of the river across from Bellevue, Nebraska—today a suburb of Omaha and the location of Fontenelle Forest Preserve. "Cat fish is verry Common and easy taken in any part of this river," the journals noted in an entry dated July 23.

On July 29, when they were north of Omaha, "we Stoped to Dine under Some high Trees near the high land" Clark wrote, and "in a fiew minits Cought three verry large *Catfish* (3) one nearly white." He noted, "Those fish are in great plenty on the Sides of the river and verry fat, a quart of Oile Came out of the Surpolous fat of one."

Although in Lewis and Clark's time there were many fish in the main channel of the Missouri, it was sometimes much easier to catch fish in quiet backwaters, as the expedition discovered. On August 15, Clark took ten men to a creek along the Missouri River where they found a beaver dam and started to fish. He wrote that "with Some Small willow & Bark we mad a Drag and haulted up the Creek, and Cought 318 fish of different kind." He listed pike, bass, perch, catfish, something he called "salmon" but was probably a brook trout, since there are no native salmon on the Missouri, "red horse," and "a kind of perch Called Silverfish, on the Ohio."

However, fishing was not a major focus of the expedition. Lewis and Clark's eyes were on the land along the river, and its wildlife, vegetation, and potential for settlement and development; and their eyes were on the surface of the river whose treacherousness demanded constant alertness. Most of the expedition's protein came from four-footed game, especially buffalo, deer, elk, and antelope. The murky waters of the Missouri made it less likely that they would see fish unless they tried to catch some, and not too many of the men seemed to be fishing enthusiasts. This is not inconsistent with the economics of development since their time. Fish can be an important source of protein, but, worldwide, fishing is a minor part of the economy.

But with European settlement along the river, catfish became a highly desired catch. They were an important food for those going west on the Sante Fe, Oregon, and Mormon trails, and they were important to the early homesteaders. In the twentieth century, catfish were fished commercially on the Missouri River and were a prized recreational catch.

Unfortunately, as it has happened in fisheries around the world, commercial fish catch on the Missouri declined in the twentieth century. Commercial catch of catfish declined steadily after World War II, decreasing by 61 percent between the 1940s and 1980s. The average size of caught catfish also decreased, meaning that overfishing had resulted in fewer adults reaching maturity. Channel catfish can grow to four feet and sixty pounds, and blue catfish can grow to one hundred pounds, but by the 1980s the channel catfish caught in the Missouri were less than seventeen inches long, and most were smaller.

In 1992 the Nebraska Game and Parks Commission closed all commercial harvest of catfish. In part this decision was to help support recreational fishing, which had ranked as the major public activity on the river.

This story is typical when habitat is destroyed and harvests are too large. Once again what happened on the Missouri River symbolizes a general pattern. Today there are more than ninety species found on the lower Missouri, including some introduced species that were not there at the time of Lewis and Clark. Of these, the pallid sturgeon is listed as endangered under the U.S. Endangered Species Act, and other species are listed as being of special concern to the U.S. Fish and Wildlife Service because of their low numbers.

Hamburg Bend Wildlife Management Area

Another restoration project is taking place at Hamburg Bend Wildlife Management Area, six miles south of Nebraska City, Nebraska, on the Nebraska shore of the river, and across from the town of Hamburg, Iowa. The Army Corps of Engineers and the Nebraska Department of Game and Parks are restoring 1,637 acres of prime hunting and fishing land and waters. Lewis and Clark passed by this part of the river in late July 1805. On July 19, near the present site of Nebraska City, the expedition passed a sandbar and found the river "wide & Shallow." Clark wrote, "Small butifull runs Come from the Plains & fall into the river." He found a deer lick and also observed that at this location, "as we approach this Great River *Platt* the Sand bars are much more noumerous than they were, and the quick & roleing Sands much more danjerous." The river was very wide, the banks soft and easily undermined by the river. There were many "gees." The next day, they passed "good land covered with Grass interspersed with Groves & Scattering timber," much as Hamburg Bend appears today where it is not farmed.

At Hamburg Bend, the Army Corps of Engineers has opened a chute—a side channel to the river meant to maintain a steady but controlled flow into the floodplain and reestablish some quiet backwaters. The current inside a chute is relatively gentle, and the water deposits sediments, creating sandbar islands. Large logs are scattered all along the chute and in the water. The created chute mimics some of the old river channel structures.

Before channelization, the Missouri River produced an incredible number of fish. The key to restoring the fish in the river is surprisingly straightforward. Many fish feed on insects, but in the swirling waters of the Missouri, insects have no place to stand. There are few boulders or places where bedrock outcrops into the riverbed. Without a foothold, insects are swept downstream. If insects have a place to stand and feed, they can become food for fish. In this constantly changing environment, snags—logs that are caught in the bottom, hung up on a sandbar, or tangled in a group—provide the main stable surface in the river. Not only do they provide a place for insects to stand, they catch prairie grasses that fall into the river—food for insects.

Boating on the Chute at Hamburg Bend. Jerry Mestl, fisheries biologist with the State of Nebraska, looks at the snags piled up on the shore, a manmade backwater that is helping restore fish on the Missouri River. *D. B. Botkin.*

Hamburg Bend and Chute Cut Through by the Army Corps of Engineers as Part of Restoration of Habitats on the Missouri River. *U.S. Army Corps of Engineers.*

The removal of snags began in the early nineteenth century, and was completed about 1950. As a result, insects became less abundant. In the unchannelized reach of the river, from Gavins Point Dam to Ponca State Park, the annual production of river insects declined between 1963 and 1993 to one-quarter of its former value.

Ironically, the snags that kill boats are also the saviors of the fish. But the snags need not be in the main navigation channel. They can lie instead only in side channels and backwaters. In this way, two uses of the river, navigation and fishery management, can coexist.

A wonderful thing about the Missouri River is the resiliency of its life, because, when the habitats are available, the river provides plenty of water and nutrients. The Missouri River is a kind of natural aquaculture. Create a few, limited kinds of backwater channels on the river, and fish production goes up. The river is so fertile, the soil so rich and well watered, that not much else is needed.

The river farms the land. It deposits fresh soil in some places, and there trees or prairies grow. Elsewhere, the river undercuts and harvests. The prairie feeds the river, but the river harvests the prairie; it harvests the floodplain forests.

Snags on the Modern Unchannelized Section of the Lower Missouri River. *James Peterson.*

A Wonderful Vista

One of the best ways to understand the lower Missouri River valley is to see it from the top of one of the nearby limestone bluffs. Lewis and Clark often climbed these bluffs on their way up the river, especially between the locations of modern-day Jefferson City, the state capital, and Columbia, Missouri. From these heights, they could read the countryside to see what its natural resources might be and judge its potential for farming, settlement, and defense.

One of the most spectacular views of the river valley is from the Les Bourgeois Winery near Rocheport, Missouri, just west of Columbia, on the top of limestone bluff. Far below are the wide river valley and the narrow, engineered main channel maintained at a minimum of nine feet deep and three hundred feet wide, cut deep enough for barge navigation. Alongside the main channel are rows of cottonwoods and willows, their bright green highlighting the levees built to protect farmland on the bottomlands beyond the channel.

The Missouri River from Bluffs near Columbia, Missouri. The river, which drains one-sixth of the lower 48 states, passes under many bridges, including the Interstate Highway Bridge just visible in the distance. *D. B. Botkin.*

"I like to bring people up here and show them the view of Overton Bottom and tell them that the Missouri River drains one-sixth of the United States, and all that water has to flow right *there*, under *that* bridge on Interstate 70," J. C. Bryant, former director of the Big Muddy National Wildlife Refuge, said as he looked down at the beautiful landscape. From the top of the Overton bluff, all appeared placid, almost gardenlike, a mosaic of bottomlands: dark soils of the few remaining active farmlands, grays of last year's weeds in abandoned farmland, stands of cottonwoods and willows greening the sands and silts.

And so, little by little, the prairie, the Missouri River with the prairie, and the wildlife and fish that once lived there, are being brought back. The mixture of agriculture and the great American prairie may actually be our future.

6

FIRE, WIND, AND WATER

THE PLATTE RIVER AND THE
LOESS HILLS WITHIN THE PRAIRIES

IN JULY 1804, Lewis and Clark crossed two ecological boundaries, one wet and one dry. They reached the mouth of the Platte River, one of the Missouri's major tributaries. And they reached the loess hills, a strange formation, unlike anything they had seen before and rare on Earth.

On July 16, Lewis and Clark were near today's Iowa–Missouri border on the river near the present location of Omaha, Nebraska. Clark wrote that the river was about two miles wide, but it was not deep because he could see snags scattered across it, and on the far shore "about 20 acres of the hill has latterly Sliped into the river above a clift of Sand Stone for about two miles." Looking beyond the river, Clark saw "a range of Ball [bald] Hills parrelel to the river & at from 3 to 6 miles distant from it, and extend[ing] as far up & Down as I Can See." He was viewing a special kind of prairie that grows on a strange and rare kind of soil, called loess. "This Prairie I call *Ball pated Prarie*," Clark wrote, seemingly because the brownish color of the soil and grasses made the hills look hairless at the top.

Our modern equivalent would be to call them monk-shaved hills, because the summits are clothed in grasses while trees and shrubs surround the bases near the water, giving the hills the rounded appearance of a monk's head with the hair shaved above the ears. The next day, Lewis rode out into this prairie and along a stream that passed through it, which he referred to by its Indian name, Neesh-nah-ba-to na. Along this stream they found "Some few trees of oake walnut & mulberry." The hills were not as barren as they looked: several of the men went out hunting and the best hunters, Drouillard, "Kill'ed 3 deer, & R Fields one," Clark wrote.

Of the few places on Earth where loess occurs, the American Midwest and China have the two largest parcels. Loess begins as silt eroded from mountains—in North America eroded by the ancient Missouri River from the Rocky Mountains and by the Platte River to the south. Rivers sort the material they carry. The faster a river flows, the heavier the material it can carry. As a result, rivers separate material they carry by size, leaving the heaviest material behind, near the mountains. A mountain stream pushes boulders during storms. Silts are fine particles and are carried a long way downstream to where the quieter waters can no longer suspend them. As

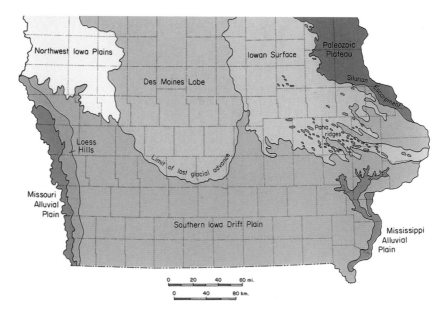

Map of the Loess Hills. Formed of a very rare kind of soil, loess hills occur in North America along a narrow strip on the east shore of the Missouri River, near present day Omaha, Nebraska. The hills were noticed immediately as unusual by Lewis and Clark. They originated from a series of geological forces each of which would appear as catastrophic if there were human observers. *From* Landforms of Iowa, *Jean C. Prior, Geological Survey Bureau, Iowa Department of Natural Resources. (University of Iowa Press: Iowa City, 1991.) p.31.*

the Missouri and Platte meandered across their floodplains over the centuries, they spilled their silt wide and deep.

A Catastrophe

Toward the end of the last ice age—between thirty thousand and seventeen thousand years ago—intense winds blew, a result of the great difference in temperatures and reflection of light between the ice to the north and the warmer bare ground to the south. The great winds lifted up the silt that the rivers had deposited, creating silt storms that piled the soil into steep silt dunes. Like sand dunes, the silt dunes formed a rolling countryside. Seen from the side, especially along a road cut, the soil is deep but not layered.

Here in Iowa, the loess sits atop glacial till—material spread by glaciers as they bulldozed across the land—and this, in turn, sits on bedrock much, much older. The bedrock of the cliffs that Lewis and Clark saw when they first viewed the loess hills was deposited in the Pennsylvanian period, more than 286 million years ago, in the age of coal formation.

Murray Hill, Within the Loess Hills Today. The steep, rugged loess hills rise above the smooth, flat surface below them. The rugged loess hills are poorer farmland than the floodplain below. *From* Landforms of Iowa, *Jean C. Prior, Geological Survey Bureau, Iowa Department of Natural Resources. (University of Iowa Press: Iowa City, 1991.) p. 8.*

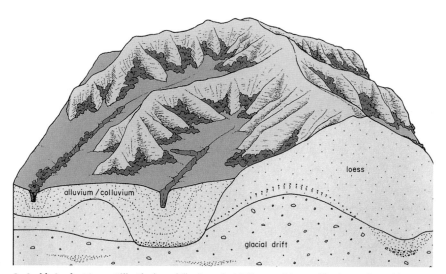

An Inside Look at Loess Hills. The loess hills sit on glacial till—material moved by the continental ice sheets, which in turn sit on hard, ancient bedrock. The soft silt that forms the loess hills erodes readily, forming steep-sided gullies. *From* Landforms of Iowa, *Jean C. Prior, Geological Survey Bureau, Iowa Department of Natural Resources. (University of Iowa Press: Iowa City, 1991.) Illustration by Patricia J. Lohmann, p. 49.*

Loess hills have their own special appeal. They evoke a certain feeling, not so much in the Greek or Roman sense of highly ordered beauty, but of nature's movement and power, temporarily at rest. Clark was affected by the loess hills scenery and the changes he could see in the prairie landscape. On July 19, three days after he first saw the loess hills, Clark went for a walk on the shore of the Missouri, after a breakfast of roasted deer ribs and coffee. He began to follow fresh tracks of elk and "after assending and passing thro a narrow Strip of wood Land, Came Suddenly into an open and bound less Prarie" where trees were "confined to the River Creeks & Small branches" and the prairie "was Covered with grass about 18 Inches or 2 feat high and contained little of any thing else."

The grass was shorter than on the tall-grass prairie that grew on other kinds of soil and had become familiar to the expedition since they had reached the Kansas River. "This prospect was So Sudden & entertaining that I forgot the object of my prosute and turned my attention to the Variety which presented themselves to my view," he wrote. He continued up a hill toward a "line of woods" where he found "a butifull Streem" that he followed for three miles to where it flowed into the Missouri River "between 2 clifts." He was near the location of modern Nebraska City, probably on Table Creek.

Loess makes poorer soil than does glacial till, partially because the soil is made up of particles of one size. The best soils in Iowa are to the east of the loess; the poorest in the loess hills along the Missouri where Lewis and Clark passed. This was one of the things that struck Clark about his bald-pated prairie: the grasses on these hills were short—about a foot and a half high—rather than tall like the grasses of the prairie he saw elsewhere, some of which reached six feet. As a result, he could get a view over a long distance to a far horizon, and that view was pleasing.

There is an irony about this landscape. It was created by ice and wind, during episodes that people, if present, would have found terrible and destructive. This seemingly peaceful landscape would not have been here without several kinds of environmental disasters, next to which the dust storms of the 1930s would have seemed like small dust devils.

Wildfires and Prairie

Another environmental catastrophe—wildfire—maintains the loess hill prairies, just as it does elsewhere in the tall grasses. Prairie fire was familiar to the expedition. On July 23, the expedition was near the Platte River mouth when Clark wrote that "Indians in this quater are in the Plains hunting the Buffalow" and that "the Praries [were] being on fire" by the Indians. And on July 20, Clark walked along the shore and found that the "Praries appears rich but much Parched with the frequent fires."

Similarly, on August 15, Sergeant Floyd—who would sicken and die within days, apparently from appendicitis—was one of three men sent out by Lewis and Clark to examine a prairie fire "at no great distance from Camp." Returning, the men reported that the fire had been set by Sioux Indians.

Historic records from other explorers of North America suggest that it was a common practice of Indians to light fires in prairie and forest. Henry Hudson reported this along the East Coast, as did other explorers. In this way, the natives of North America had a large effect on the landscape, often converting it, with fire, to a condition more useful to them, and also more useful and beautiful to the Europeans.

The next spring, on April 9, 1905 the day that the explorers left their winter fort at the Mandan villages, Lewis noted that "the Bluffs of the river which we passed today were upwards of a hundred feet high . . . [and] very broken and many of them have the apearance of having been on fire at some former period." And the next day Lewis wrote, "The country on both sides of the missouri from the tops of the river hills, is one continued level fertile plain as far as the eye can reach, in which there is not even a solitary tree or shrub to be seen except such as from their moist situations or the steep declivities of hills are sheltered from the ravages of the fire." More-over, "about 11/2 miles down this bluff from this point, the bluff is now on fire and throws out considerable quantities of smoke which has a strong sulphurious smell"—probably a fire in coal near the surface, where some-times methane burns.

The same ideas that formed Jefferson's view of what the American West would be—a land of balance, constancy, and harmony—came down to the twentieth-century idea that fires are unnatural and bad. The U.S. Forest Service's Smokey Bear advertisements reflected this belief. With the sup-pression of fire in the twentieth century, the vegetation on the loess hills and the surrounding prairies—as well as many other ecoregions in North America—changed.

In the prairies, fire-suppressed areas typically converted to woodlands if rain and snowfall were sufficient. In Iowa, within the loess hills and sur-rounding prairies, the eastern red cedar, a small tree that grows approxi-mately twenty feet high, is one of the first pioneer trees to invade the prairie grasses. Red cedar grows rapidly in bright sunlight, when it is not shaded by taller trees. Seeds of red cedar are eaten and excreted by birds and animals, and the seeds are spread widely. If the seeds fall on bare soil, they can ger-minate and survive. But if there is a dense cover of grasses, then the seeds land within the grass and do not reach the soil.

So there are two major intervals following a fire when red cedar can invade: soon after a fire, before the grasses are well established; and ten or twenty years later, after the grasses have matured and some have died back and exposed bare soil. Where rainfall is high enough, cedar can survive past the seedling stage and grow for forty or fifty years. Without fire, in this east-ern edge of the prairie, in locations where the rainfall is relatively high, the cedars grow well.

Loess Hills and Floodplain Topography. The loess hills are visible from space, as shown by the satellite image from 140,000 feet above the Earth's surface. This is a color-infrared photo—green vegetation appears as red. The area shown is along the meandering Big Sioux River near present-day Sioux City, Iowa. The heavily cultivated floodplain shows a sharp boundary with the Loess Hills region, which have, in contrast, a dense network of streams among steep hills. The land use on the loess is highly variable, in contrast to the surrounding floodplain, which is almost all in crops. This photo was taken in April 1980 approximately seventeen miles northwest of Sioux City. © *Gary Hightshoe.*

One place to see the effects of fire suppression on a prairie is Iowa's Turin Loess Hills Wildlife Management Area, where experiments are conducted to test the role of fire in maintaining tall-grass prairie. There, a view across the rolling hills and valleys shows unburned hills that are dark green from a dense cover of red cedar. Grasses include big bluestem, prairie grass, Indian grass, foxtail grass, and little bluestem, a prairie grass common in the eastern United States.

Unburned hills have a distinct pattern. On the south slopes, the vegetation is almost all prairie grasses, while the north slopes are heavily wooded. As a result, when one looks south and sees only north slopes, one sees only a wooded countryside. Without another view, you would believe you were in forestland. But turning north, one sees only south slopes, which are hill after hill of brown grasslands. It is the clearest evidence that the direction of a hillslope affects its vegetation. A south-facing slope in the Northern Hemisphere gets sun most of the day, and sun dries out the soil. A north-facing slope is shaded a good part of the day, stays cooler, and

Turin Loess Hills Wildlife Management Area, Iowa, with Dahlias and Red Cedar. Red cedars come into prairie when fires are suppressed. They grow in dense stands on the north slopes, which are cooler and have moister soils. Heavily wooden north slopes are visible in the distance while red cedars form a savannah on the ridge tops and south slopes, as in the foreground. *D. B. Botkin.*

retains its moisture. The steep, short hills formed by loess accentuated these differences.

The drier south-facing slopes are more likely to burn. In this way, a south slope is like landscape far to the west, while a north slope is like land to the east.

Although the Indians lit fires, they were, of course, not the only source of fire. One of the more peculiar sources of fire was discovered on August 24, by Clark. On this day, the expedition was near the site of modern-day Ponca State Park, in Nebraska. Lewis and Clark saw "rugged Bluffs" on the southwestern shore of the river rising "about 180 or 190 feet high." These bluffs are northeast of present-day Newcastle, Nebraska, and six miles southeast of Vermillion, South Dakota. Today they are known locally as the "Ionia Volcano." Clark wrote that the bluffs had "been lately on fire and [were] yet verry Hott." They examined these rocks and tried to ascertain what they were, referring to them as having a "Great appearance of Coal & imence quantities of Cabalt in Side of that part of the Bluff which Sliped in."

Like the vegetation on the loess hills, most prairie land—and most temperate and warm-climate grasslands around the world—require fire. But the frequency of fire required to maintain grasslands varies with precipitation and temperature. In the American West of Lewis and Clark, the great prairies developed in the rain shadow formed by the Rocky Mountains, driest just east of these mountains and becoming wetter to the east. The average rainfall is lowest just east of the Rockies—in the high plains a hundred miles east of Denver, for example, it averages twelve to sixteen inches a year. Then rainfall increases as one moves eastward, to twenty inches per year at Dodge City; twenty-eight inches near Lincoln, Nebraska; and thirty-six inches east of Kansas City. This is the primary reason that the eastern prairie is most likely and rapid to convert to forest if it does not burn, while the westernmost prairie converts to a desert scrubland.

Grazing by buffalo, elk, pronghorn, and deer, as well as feeding by prairie dogs and other small herbivorous mammals, tends to maintain the prairies as well. But each of the grazers finds species of plants that are unpalatable to it. The animals do not eat those plants, and, without any fire, the prairie converts to an area dominated by them—a combination of what people call "rank weeds" that are less beautiful to us than the grazed and burned prairie land.

Erosion and the Loess Hills

Although loess soil is not the most fertile, it is easily plowed and free of rocks and boulders, and was rapidly converted to agriculture following the time of Lewis and Clark. But just as loess is the product of catastrophes— fire, wind, and glaciers—it also falls victim to human-induced ones. Loess erodes easily, and plowing of it, especially by methods used in the nineteenth and much of the twentieth century, led to gulleys and soil loss.

Agriculture made loess more vulnerable to erosion in another, indirect way. Water filters comparatively quickly through the loess soils, and the upper slopes of the loess hills are dry compared to the land along the river floodplain. When fires were common and the land was not subjected to the plow, the hills were covered by drought-tolerant vegetation, especially on the southwestern slopes. These plants intercept the intense and drying westerly winds. Crops planted on these hills were poorer at anchoring the soil, so the loess, already made vulnerable by the plow, was less protected by its new, introduced vegetation.

Soil Slumping in the Loess Hills. The loess hills are soft and erode easily, especially when plowed and planted in crops. Here a section of a loess hill has broken away, probably as a result of plowing. *D. B. Botkin.*

The loss of loess took place during a time of great human alterations of the tall-grass prairies—beginning in the second half of the nineteenth century and continuing since. In Iowa, before European settlement, twenty-eight million acres of prairie spread over the landscape, containing two hundred species of plants, especially bluestem, dropseed, compass plants, coneflowers, and gentians. Today these are fields of corn, wheat, oats, hay, and pasture. Prairie wetlands were once common, but most have been drained by farmers to provide more cropland. At first the wetlands were drained by wooden pipes, then wooden pipes were replaced by tile ones, and more and more wetlands were drained, eliminating much wildlife habitat. During the same time, Iowa lost much of its forestland. This land, most of it just beyond the vision of Lewis and Clark to the north and east, occupied 19 percent of the land area of what became that state. That northeastern part of Iowa where the landscape was more heavily wooded was dominated by white oaks and sugar maples, as well as the red cedar, which grew especially well on limestone. Today's valuable black walnut also grew abundantly in certain habitats.

The forests were cut to provide fuel for steamboats on the Mississippi and Missouri rivers and, beginning in 1855, to help build the railroads. Fuel was required for steam engines and lumber for ties on ten thousand miles of track; timber was harvested at a rate of about eight hundred trees and six acres of forest for each mile. The ties lasted no more than ten years, requiring more deforestation. But much of the loss of Iowa's forests came in the twentieth century: only four million of the original nineteen million acres were lost by 1900.

With the change from prairie and forest to agriculture, wildlife habitats declined. Much of this took place also beyond the vision of Lewis and Clark, in the north of Iowa away from the Missouri River. This heavily glaciated landscape was a mosaic of prairie uplands and prairie marshes that provided habitat for ducks, geese, trumpeter swans, whooping cranes, marsh wrens, and yellow-headed blackbirds, along with muskrats, turtles, dragonflies, fish, frogs, and salamanders. Hunting also decreased wildlife populations, as the first European settlers found the deer, turkeys, and prairie chickens easy targets. According to a report on the state of Iowa, "people gathered duck, goose, and swan eggs in the spring and shot the birds virtually year-round for food and feathers. Market hunters also slaughtered shorebirds and waterfowl by the hundreds, often shipping the birds to restaurants in eastern cities. River otters and beavers initially thrived in most rivers, streams, and marshes, and trappers sought them for fur during the heyday of the fur trade in the late 1700s. During the nineteenth century, trapping pressure, habitat loss, water pollution, wetland drainage, and stream channelization gradually took their toll. Beavers and otters were essentially gone from Iowa around 1900."

Fire was not envisioned as part of the great balance of nature by the mapmaker Arrowsmith, whose map was in Lewis's kit, nor by Jefferson. However, ecologists have found fire to be an essential, if violent, characteristic of that vast, beautiful, and wondrous sea of grass that Lewis and Clark saw, the one that so distracted the otherwise practical Clark and charmed, while challenging the pioneers who crossed the prairies after them. Their settlements and farms rapidly—amazingly rapidly—eliminated that beautiful landscape.

The Platte River Meets the Missouri

Just as the loess hills provided a novel countryside to the expedition, the Platte River, flowing into the Missouri River not far south of the loess hills, was a radical departure from the river the expedition had been following. On July 21, Lewis and Clark had reached the mouth of the Platte, already at that time a well-known tributary of the Missouri. It was a big river, one of the largest of the Missouri's tributaries, and it was fast and full of large, sandy sediment. The sand was "remarkably small and light" and "easily boled up and . . . hurried by this impetuous torrent in large masses from place to place in with irristable forse," Lewis wrote. If he needed any more evidence that rivers, like the rest of nature, are in constant flux, he found it in the Platte, which was even more turbulent than the Missouri. In the course of a few hours the sands collected and formed sandbars that "as suddingly disapated to form others and give place perhaps to the deepest channel of the river."

The Platte flowed, Lewis wrote, with "a boiling motion" that he speculated was the result of "the roling and irregular motion of the sand of which its bed is entirely composed"—a good insight. The bottom of the Platte, like the Missouri, is formed of its own sediment rather than bedrock or large boulders. These bottom sediments are moved along by the water much the same way that wind drives sand. Sediment dunes form from water currents just as sand dunes form from the wind. These little hillocks force the water up and over, creating turbulence. The resulting motion flows downstream and upwards, reaching the surface as large bubbling bursts that look just like boiling water in a kettle. You can see this motion in the Missouri River today when you stand along the shore.

Lewis and Clark measured the speed of the rivers and found the Platte the fastest, running "at least" eight miles an hour. The Missouri above the Platte was running about three and a half miles an hour in its widely dispersed, complex floodplain of many channels. Below the Platte and influenced by that river, the Missouri was running five and a half miles an hour. The faster a river flows, the heavier and coarser material it can carry, so the Platte could carry a sandy load, while the Missouri could lift only smaller silts. Also the two rivers drained different kinds of countryside. The Platte watershed is mostly sandy soils eroded from the Rocky Mountains and then deposited in western Nebraska long ago, only to be picked up again and transported by the Platte. The Missouri drains a landscape where the ice-age glaciers created soils of many sizes of particles, including small silts and clays.

At the mouth of the Platte, the two kinds of sediment loads came together but did not yet mix. The Platte was so swift and powerful that its current, having entered the Missouri River channel, forced the Missouri's waters against the far bank "where it is compressed within a channel less than one third of the width it had just before occupied," Lewis wrote. The currents could be distinguished by their colors. The Platte did "not furnish the missouri with it's colouring matter as has been asserted by some, but it throws into it immence quantities of sand." Lewis claimed that the separation of the currents "abates but little untill it's junction with the Mississippy."

Both rivers were so full of sediment that neither was clear, but there was a difference. According to Lewis, the Platte deposited "very fine particles of white sand while . . . the Missoury is composed principally of a dark rich loam—in much greater quantity." The Platte's water was "turbid at all seasons of the year but is by no means as much so as that of the Missourie." These two painters of landscape had different palates and different styles of painting.

The Platte has a great effect on the Missouri in other ways as well. Above the Platte and below Gavins Point Dam—the farthest downstream of the Missouri's dams—the channel is eroding and deepening, but this erosion takes place only in the main river channel, not in the surrounding flood-plain. The river is incising itself into the engineered floodplain.

Below the mouth of the Platte, the river is depositing material on the bottomlands. The floodplain is accumulating material, while the main channel, functioning as designed by the Army Corps of Engineers, is not changing its depth: it is neither eroding nor sedimenting.

The accumulating material on the Missouri River floodplain below the Platte has an important effect on floods. This accretion is raising the level of the floodplain relative to the artificial levees, shrinking the height of the levees relative to the floodplain. The total volume of water that can be held back by the levees is therefore less. It is like a stream carrying sand and silt into a swimming pool and depositing those sediments in the quiet waters. Over time, the sediments build up on the bottom, and the pool can hold less water. This was a factor that made the large floods of the 1990s so cata-strophic. In 1998, the Army Corps of Engineers reduced the flow from Gavins Point Dam, claiming that bottom aggradation below Omaha was causing lowland flooding.

The aggrading land below the Platte has other biological effects. As the floodplain builds up, its soils become drier; there are fewer wetlands but the soil becomes better for farming, at least where the deposits are not heavy sands. Overall, the floodplain habitats become fewer and simpler.

The Missouri at the mouth of the Platte is revealed as a complex system: fingers of tributaries feed waters of different colors, qualities, speeds, and amounts into the main channel. In this sense, the Missouri is many rivers, not one. What happens on the tributaries affects the Missouri, and many of these tributaries have been channelized and controlled to prevent flooding. Water from channelized tributaries flows faster and adds more quickly to the floodwaters of a rising and dangerous Missouri. The quality of the water from tributaries affects the Missouri, as does the chemical runoff from farms, industries, and houses. Just as the Missouri drains one-sixth of the continental United States, so does that land affect the quality of the Missouri's waters.

Today the mouth of the Platte River is easily reached within the Shilling Wildlife Area of Nebraska. There you can see the tumbling, mixing currents of the Platte and the Missouri as these still rush together, boil, and evoke in us the feel of prairie and river countryside, reminding us of the intimate link between rivers and their landscape, and the constancy of change in nature.

Along the Platte River

On July 21, after viewing the mouth of the Platte River and measuring the flows of the Platte and the Missouri, Lewis and Clark traveled a ways up the Platte. They found it shallow—not more than six or seven feet deep anywhere. One of the men of the expedition had spent two winters along the Platte and told Clark that the Platte "does not rise 7 feet, but Spreds over 3 miles at Some places." Later pioneers would say the Platte was "a mile wide and an inch deep."

The pioneers followed the Platte. Although Lewis and Clark—as well as Jefferson—hoped and assumed that the Missouri was the best way west, expecting there to be an easy portage between its headwaters and the Columbia River, in fact the better route was along the Platte. Ironically, if the Lewis and Clark expedition had headed west at this location, they would

The Platte River. *D. B. Botkin.*

have had an easier trip to the Pacific coast. The Platte became the river of the Oregon, California, and Mormon trails. Lewis and Clark could not have considered an alternative to the Missouri because Jefferson's instructions to them were explicit: follow the Missouri River to its source. And there were no data to tell them that the Platte would provide an easier route.

Since it was assumed that nature was symmetrical—the Missouri's source must be in mountains no more difficult to cross than the Appalachians—their route would have seemed to be the best one. If they had had access to aerial photos and satellite images, they would have known that the path along the Platte was much to be preferred. But to their credit, confronted daily with nature's variations and lack of symmetry, they responded to what they did observe with understanding, common sense, and wisdom.

The Platte Today

Today the Platte is worth seeing as part of a Lewis and Clark rediscovery trip, because it is one of the few major rivers of America that has not been greatly altered by channelization and dams. It is an interesting river to follow. Many roads cross the Platte through central Nebraska, and it is possible to find a place along one of these to stop and see the river. Some Nebraska state parks are on the Platte, but few have direct access to the river.

Although the North Platte is dammed and much water has been removed for irrigation (the river no longer flows with the power that Lewis and Clark saw), the Platte is one of the least altered of the Missouri's major tributaries. If you want to see a river that resembles the original lower Missouri before its channelization and before its levees, one of the best things you can do is travel along the Platte. And the Platte River's Big Bend in the center of Nebraska is one of the most important habitats for migratory birds. More than 240 bird species have been seen there, including sandhill cranes, whooping cranes, piping plovers, and the least tern, species listed as endangered or of special concern.

For those interested in wildlife, Grand Island, a major Nebraska town, is a must-see because it is a major stopping point for migrating birds, especially sandhill cranes. In the spring and fall, more than five hundred thousand of these large and magnificent birds stop for four to six weeks to feed and rest before continuing to their breeding grounds in Canada, Alaska, and Siberia or south to wintering ground. Lewis and Clark saw the cranes during their trip up the Missouri, and these birds are characteristic of the central United States. Their wingspan reaches six and a half feet, making them one of the largest birds of North America. There are several parks at Grand Island that are near or on the river, and from these you can see the incredible abundance of the sandhill crane.

The sandhill crane is a success in terms of conservation. It was the first bird ever protected by an international treaty, the 1916 bird migration treaty between the United States and Canada. Protected from hunting, this species proliferated so much that in the 1960s, farmers found these birds were a pest as they migrated in huge numbers and ate grain along the way. Hunting was proposed as a solution, but this bird has never attracted a great

The Whooping Crane, an endangered species, uses areas along the Platte River during its migration. The critical habitat areas on the Platte for this species are the Central Platte River from Lexington to Shelton, Nebraska. The whooping crane is America's tallest bird, between four and five feet tall. Its number plunged to fourteen in the late 1930s, and now, through careful protection, it numbers about two hundred. At the time of Lewis and Clark, the whooping crane nested from central Illinois to eastern North Dakota and north through the Canadian prairie provinces. Today it nests only in far northern Canada, at Wood Buffalo National Park. The entire population migrates as a unit to the sole remaining wintering ground, Aransas Wildlife Area, Texas. *U.S. Fish and Wildlife Service/photo by Luther Goldman.*

The Pallid Sturgeon, a unique fish of the Mississippi–Missouri system, including the Platte River. It is a fish of ancient heritage, like its relatives, the other sturgeons. This fish grows to five feet long and eighty-five pounds. Today it is rare throughout the former range. It is one of the strangest animals along the Lewis and Clark trail. *U.S. Fish and Wildlife Service.*

many hunters. Once numbering only 15,000 or 20,000, it has expanded greatly, repopulating the prairie.

The Platte River is also considered critical habitat for several species listed as endangered or threatened under the Federal Endangered Species Act: the whooping crane, piping plover, interior least tern, and pallid sturgeon.

The sandhill crane, the sandy floodplain, and the Platte characterize prairie countryside that Lewis and Clark traveled through for many months.

Today the Platte is an odd, shallow, sometimes dried-out looking waif of a river meandering along a wide and sandy floodplain where scattered and usually half-starved looking cottonwoods struggle to survive or, having lost the battle, lie as snags in its shallow current. Between towns, the Platte and its countryside have a lonesome, open feel that is part of the character of the American West—open country, open sky, dry land.

It was the shallowness of the Platte that probably saved it from channel-ization and dams. "An inch deep"—or even the real average depth of the Platte—isn't enough for a steamboat. The Platte became the route west for those in Conestoga wagons, on horseback, or on foot, guiding the way and giving enough water to drink but never enough for navigation.

But the Platte has changed: Its waters were first diverted in 1838, and by 1885 the demand for Platte water for irrigation exceeded the average flow. There are thousands of diversion structures, and nearly 70 percent of the flow is used. The Platte is a river much closer to being an inch deep today than it was at the time of Lewis and Clark. Present flows on the Platte are about one million acre-feet a year, 4 percent of the water that flows down the Missouri River. As a result of water diversions, the proportion of open areas, wetlands, and forested floodplains has changed, leading to concern about whether enough bird habitat remains.

The state of Nebraska has a fine series of parks along the Platte River, but most of them are designed for recreation away from the water. One of the best places to see the Platte River is at Platte River State Park in Louisville, about halfway between the cities of Lincoln and Omaha. East of Lincoln, Nebraska, State Route 50 crosses the Platte River at Louisville, Nebraska (reached from Lincoln by car via Interstate 80 to State Route 370). Here you can view the modern Platte, with its snags, floating logs, islands, and wetlands along the shore. And here the Platte remains heavy with sediment—so heavy that, most of the time, one cannot see into the water at all. Although the Platte River is not channelized, its banks not altered by levees, it is much altered by modern land use. At this crossing of the Platte, one can see across the river a large rock quarry where big trucks stir dust into the air.

The Platte River has been largely ignored as a recreational river, ignored for its qualities as a great prairie river. Louisville State Recreation Area is adjacent to the river, but ironically, the quarry buildings are visible but the Platte itself can just barely be seen through cottonwoods, so that the river is not in a scenic part of the park.

Similarly, the Platte is virtually overlooked at Mahoney State Park, Nebraska, a beautiful facility with many kinds of outdoor recreation and a pretty lodge. The lodge is the one place where one can see the river, but that building imposes itself between the river and everything else. One can only view the Platte from inside the lodge or on its porch.

It is shocking that the Platte River, one of the major rivers of the Great Plains, is so hard to reach and admire. But this is how our society has approached nature throughout much of the twentieth century: as something very separate from us.

One of the great ironies of the natural history of the Lewis and Clark travels is that some of the countryside that most impressed them, and that would impress us as well, either no longer exists, or has been altered, or is obscured from view, even in parks whose purpose is to provide recreation and relief from modern labors. Of the two dominant features of this part of Lewis and Clark's travels, the loess hills are more celebrated than the Platte River. There are numerous loess hills nature preserves, and the state of Iowa has created a travel route called the "Loess Hills Scenic Byway." This is 220 miles of paved roads more or less paralleling Interstate 29 that travels north-south near the Missouri River and provides designated roads through the scenic Loess Hills region of western Iowa.

In contrast, the Platte is mostly ignored or hidden from view by modern land uses. The major exception is Grand Isle, where viewing of the annual migration of sandhill cranes has become a well-known attraction. Otherwise, one has to hunt carefully for a view of a landscape that once covered a vast area of the United States: the prairie in its various forms—tall grass, loess hills, mid-grass, and short grass, and the great prairie river, the Platte. A land-scape that could evoke near-poetry from many early travelers is lost to us.

Efforts to restore prairie are meager compared with efforts to conserve scenery and living resources along the rest of the Lewis and Clark trail—the

The Sandhill Crane. *U.S. Fish and Wildlife Service/photo by John and Karen Hollingsworth.*

Rocky Mountains, the Columbia River system, and the forests of the Pacific Northwest. These deserve the attention they receive, but it would be a greater America that also chose to restore large areas of the once-great American prairie and its rivers. Perhaps if we did this, the vast region of middle America, sometimes referred to as a land that is losing its human population and often referred to as an aesthetic wasteland, would thrive again.

The Unchannelized Missouri

The Missouri River is channelized up to Gavins Point Dam on the Nebraska-Iowa border. For a few years, the paddlewheel boat *Far West* made a tourist trip from Vermillion, South Dakota, north near to the base on the dam. It was a beautiful trip and one of the few ways a traveler without his own boat could get onto the lower Missouri. From Sioux City north to the dam, the river is heavily used for recreation during the summer. But among the things that take away from the beauty is a collection of junked cars put on the riverbank to stabilize it. Perhaps at one time, when rivers were perceived simply as ways of transportation, including transportation of garbage and trash away from town, this use of junked cars solved two problems: it got rid of the junked car, and it stabilized the riverbank. That the cars contained lead and other toxic heavy metals was not considered a problem. Since the river was not valued for its beauty, the attitude seems to have been "Who cares?" Today the answer is "We care." But the cars still lie on the banks of the Missouri.

Near Gavins Point Dam, it is not hard to imagine how the Missouri River appeared to Lewis and Clark. One needs only to go to Nebraska's Ponca State Park, just downstream a way to the wild and scenic stretch of the Missouri River. This park marks the southern limit of the unchannelized and undammed stretch of the Missouri. It is the most accessible location to experience this reach of the river and its nearby forests, the river and landscape as Lewis and Clark knew it to be.

Ponca State Park is site important for conserving endangered species, as a sign posted by the Army Corps of Engineers warns: "Attention boaters and recreationists; least terns and piping plovers are protected by State and Federal Endangered Species Laws. Both species nest on sandbars and beaches on the river. Some of these nesting areas are posted as closed to all access. Do not disturb these birds."

Early in 1998, a new overlook was built on a bluff in the park. It was a state and national park service project, and it contains four excellent interpretive signs. Three states—Nebraska, South Dakota, and Iowa—are visible from these bluffs.

The main road into the park passes through pleasant, shady forests to a picnic area along the shore where you can see floating and half-sunken logs interspersed with sandbars, and the river meandering and cutting away at the bank. Steep, almost vertical cliffs next to the river are pock-marked with swallow's nests—similar to ones either mentioned by Lewis and Clark when they passed this way. They are generally light colored but have bands shading into dark, almost coal-like rocks.

Cottonwoods, willows, and ash line the shore. On a sunny summer day, the setting is a quiet and peaceful, pastoral scene. Breezes rustle cottonwoods, but there are few other sounds—a solitude unlike much of the lower Missouri River, crowded in by highways and railroads, industries and

A Steamboat Ride on the Missouri River near Gavins Point Dam. For several years in the 1990s, the steamboat, the *Far West*, provided a wonderful tourist ride up to Gavins Point Dam, but economics were against it: the season was too short for the boat owners to make a profit or break even. While the *Far West* sailed here, it provided one of the few commercial rides of the time on the Missouri River. *D. B. Botkin.*

cities, for most of its length downstream, its surface resonating with the sound of motorboats.

Meandering roads within the park lead to a series of low bluffs, wooded with eastern deciduous forest vegetation: eastern white and red oaks, eastern red juniper, basswood, and dogwood descend steeply to the shore. It is among the richest woodlands one can see on a journey into Lewis and Clark country. These woodlands have a rich, dark surface of humus and leaf and twig litter; they are dense with shrubs and saplings, and the soil is rich and organic, with its own scent. Rubbing a little soil between his thumb and first finger, an experienced naturalist can tell much about its fertility from its aroma and from the feel of the slickness of fine silts and clays, the grittiness of a little sand, and the softness and pliability of leaves and twigs. This handling of the soil, a standard practice among field soil scientists, is one of nature's communications with us, a kind of braille. It might seem crude, but if one does it enough one learns to tell one kind of soil from another quickly.

The following spring, when he was in western North Dakota near the Little Missouri River, Lewis described the scents of the landscape when he wrote on April 14, 1805 "the upland is extreemly broken, chonsisting of high gaulded nobs. . . . on these hills many aromatic herbs are seen; resembling

Junked Automobiles Used to Stabilize the Missouri River Bank near Gavins Point Dam. A strange piece of scenery near the upriver end of the *Far West*'s trip. *D. B. Botkin.*

The Way It Was: The Lower Wild and Scenic Stretch of the Missouri River, at Ponca State Park, Nebraska. *D. B. Botkin.*

in taste, smel and apperance, the sage, hysop, wormwood, southernwood, and two other herbs which are strangers to me; the one resembling the camphor in taste and smell" and another "of an agreeable smel and flavor; of this last the Atelope is very fond; they feed on it, and perfume the hair of their foreheads and necks with it by rubing against it."

Sights, sounds, scents, the feel of rich soil—all are part of observing nature. To know the nature of the American Midwest, the nature of the river, the prairie, and the forests, is more than to see it and pass it by. Lewis and Clark reached another level of sensory experience, one that sometimes manifests as an intuitive knowledge or understanding. This is how to read nature, to learn nature's stories.

7

WINTER ON THE PLAINS: LEWIS AND CLARK AMONG THE MANDANS

O N O C T O B E R 21, Lewis and Clark reached the Hidatsa and Mandan villages where they would spend their first winter—and from which, the next spring, they would send back live prairie dogs. Auspiciously, the day they arrived, they saw a beautiful plain "Covered with Buffalow," one of which they shot for food. Although winter was approaching, their arrival went pretty much as planned. The maps that they had seen in St. Louis in 1803 and early 1804, and copied and carried with them, showed the Missouri up to these villages. They were still in "discovered" country, land that had become known to Europeans, mainly French-Canadian traders, trappers, and hunters; the usual raggle-taggle of the curious, the rough, the ill-fit, and the outcasts of civilization. From such as these, the Mandan and Hidatsa villages were known to Lewis and Clark as a place of peaceful Indians where they might overwinter.

Winter on the Plains

Winter had approached the expedition with a deliberateness recorded firmly in the pages of their journals. Methodical as always, Lewis and Clark had reached this destination just before winter set in: they knew that it would be impractical to proceed farther that year. Observant of everything and determined in their attempts to record what they saw, Lewis and Clark wrote of the change in the seasons as they approached the area that is now near Bismarck, North Dakota. On the first of October, the leaves of ash and "popular" and most of the shrubs had begun "to turn yellow and decline." There was a "slight white frost" on October 4, and they saw brant and geese "passing to South"; frost the next night; and teal, gulls, and mallards. A week later, the cottonwoods were yellow and their leaves falling; on the fourteenth, the leaves of the ash, the elm, and all other trees except cotton-wood had fallen; and on the seventeenth, snow geese passed overhead. Pronghorn antelope passed on their fall migration.

Soon after arriving Lewis and Clark began to build their cabins and a fort, and to meet with the local Indians. Several closely related groups of Indians lived in the area in small villages, using different if related languages. Lewis and Clark visited the nearby villages, including Awatixa Xi'e village on the Knife River, where they spoke with the chief. Today, this location, called the Knife River village is a National Historic Site, historic because Indians lived there at the time of European expansion into the American West. It also is a significant archaeological site, because Indians inhabited this village for about 10,000 years. Its villagers were Hidatsa Indians known as the Awaxawi, close relatives of the Mandans. Charbonneau and his wife Sacagewea probably lived in this village before joining the expedition.

We usually think of natural history as consisting of nature without peo-ple, but the American West as seen by Lewis and Clark was a vast ecological system that included, involved, and was affected by people. The people who lived within that ecological region depended on its characteristics and adjusted their activities, behavior, and societies to survive within it. The more stressful the environment, the easier it is to see this connection between people and nature. Winter with the Mandans and Hidatsa, within the dry countryside west of the 100th meridian—the longitude generally considered to mark the boundary into the short-grass prairie—reveals these nature–people connections.

A Center for Trade

The Mandan, Hidatsa, and Arikara Indians were famous among the other Indian tribes as traders. According to the National Park Service, which maintains the Knife River Indian restoration at Awatixa Xi'e Village, the Indians who lived there traded crops, meat, and hides, as well as Knife River flint, used widely for tools and weapons. The Mandan, Hidatsa, and Arikara Indians led a settled, agricultural life, which allowed them to trade the products of their agriculture with the nomadic tribes. They made use of the land for farming and its location for trade, since they were near the Missouri River, a major transportation route. Goods that passed through these villages originated from all parts of the continent. Shells, including those used as wampum, came from islands off the California coast near Santa Barbara; native copper from Great Lakes Indians came from Isle Royale, the large island in northwestern Lake Superior; and obsidian came from nearby Wyoming.

Europeans traveling through this area joined the trade. According to the National Park Service, there is evidence that European goods arrived before the Europeans themselves—early seventeenth-century deposits of glass beads and fragments of iron were found at the Hidatsa and Awatixa Xi'e villages.

Karl Bodmer, the famous landscape artist, made watercolors of the villages, village life, and many of the Indians when he accompanied Prince Maximilian of Germany in retracing the travels of Lewis and Clark in 1833 and 1834. Living in the days before photography and being a person of considerable means, Maximilian brought Bodmer along as his personal illustrator. Bodmer's scenes of these villages, show as many as one hundred and twenty round lodges, each home to an extended family of perhaps ten to thirty people. One of these villages could have had a population numbering between 1,200 and 3,600.

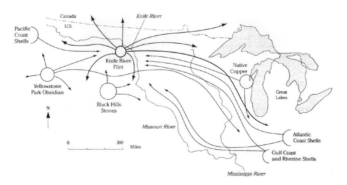

Some Trading Routes to and from the Knife River–Mandan Villages.
Knife River Indian Villages National Historic Site.

Clark observed on October 27, when he visited one of the villages and wrote that "the houses are round and Verry large Containing Several families." The dwelling sites were crowded so close together that there was little room between them even to dry corn, suggesting either a close-knit social structure or a need for protection against other tribes. There were fifty-one such depressions visible. At least four hundred people occupied this village from the late 1790s until 1834.

The Mandan and Hidatsa villages were somewhat away from the main channel and up on the prairie. Lewis and Clark chose to build their fort near a Mandan village but also down on the floodplain of the Missouri River. The Indian villages' upland setting provided good defense, access to water, and good cropland. The Indians also had winter settlements on bottomlands protected from the winds where there was easy access to cottonwoods and other trees for firewood. Both settlements took the characteristics of the landscape into account and were in tune with it, much as prairie dog villages are located on soils and near vegetation suited to them. The explorers placed their cabins just below wooded bluffs that provided protection from the winds. People fit into the landscape, and therefore into nature, by observing their surroundings and creating homes and villages in locations close to important resources. This is a lesson our modern technology has let us forget, but it is one that is still true.

In our day, we are used to seeing houses developed in large tracts where the land is bulldozed and flattened, where water is supplied by a central water authority, trash is picked up, sewage systems take away other wastes, and central heating and air conditioning keep us comfortable. Within this modern environment, it is easy to think that there is no connection between the location of a settlement—village, town, fortification, or city—and its environment. But that wasn't true in the past, and it isn't really true today.

Prairie Dogs

Among the many animals Lewis and Clark met on their journey two of the most intriguing are the prairie dog and the black-footed ferret, which they saw in the ecological region where the Mandans lived. They seemed to have thought so as well. On April 3, 1805, Lewis and Clark were preparing to leave the Mandan village where they had spent the winter. They were pack-

George Catlin (1796–1872). *Hidatsa Village, Earth-Covered Lodges, on the Knife River*, **1810. (1832).**
Oil on fabric: canvas mounted on aluminum, 28.5 x 36.6 cm. *Smithsonian American Art Museum, Washington, DC / Art Resource, NY.*

Aerial View of Big Hidatsa Village. *North Dakota Department of Transportation.*

ing samples that would be taken back to President Jefferson by the soldiers who had accompanied the expedition to this location. In his journal, for example, Clark reported that Cage 6 "Contains a liveing burrowing Squirel of the praries." Lewis and Clark sent the president a prairie dog, which actually reached the East Coast alive. In retrospect, it was a telling animal to send, not only because the prairie dog was unknown to European civilization and science, but also because human beings and prairie dogs are both social animals and both live in villages. Both construct elaborate homes and pathways between their homes (in the case of the prairie dog, underground tunnels). Among the social mammals, it may not be an overstatement to say that prairie dogs and people build the largest "cities." At the time of Lewis and Clark, prairie dog villages covered large areas and were home to a great number of animals. Even at the turn of the twentieth century, a single, immense prairie dog town in Texas, studied by scientists, covered an area 100 by 250 miles and was estimated to contain four hundred million prairie dogs.

Lewis and Clark had first seen prairie dogs when they went out walking together on September 7, 1804—a month before they reached the Mandan

A Prairie Dog. *U.S. Fish and Wildlife Service/photo by Hans Stuart.*

villages, but when they were in the short-grass prairie. They had climbed a domelike hill about seventy feet high and on their way down, Clark wrote, "discovered a Village of Small animals that burrow in the grown" and that they estimated covered about four acres, with "great numbers of holes on the top of which those little animals Set erect make a Whistleing noise and whin allarmed Slip into their hole." Lewis and Clark poured a lot of water—more than five barrels—into a prairie dog hole and flushed one out, killed it, and examined it.

"Those Animals are about the Size of a Small Squrel," Clark recorded, "& thicker, the head much resembling a Squirel in every respect, except the ears which is Shorter, his tail like a ground Squirel which thy Shake & whistle when allarmd." Clark also noted that "the mouth resemble the rabit, head longer, legs short, & toe nails long ther tail like a g[round] Squirel which they Shake and make chattering noise ther eyes like a dog, their colour is Gray and Skin contains Soft fur."

These small, social animals were familiar to the plains Indians and to the Canadians. Clark noted that they were called by the French "pitite Chien," or little dog. Lewis later wrote a more complete description that was the first scientific description of these animals.

At one time, prairie dogs were widespread throughout the short-grass prairie, the western, drier prairie countryside that does not support the grand tall grasses of the eastern plains. These animals feed on grasses but need to see from their burrows to protect themselves and would not do well in the tall-grass prairie where they could not see beyond the next clump of grass. At the time of Lewis and Clark, short-grass prairie extended north-south from southern Saskatchewan, Canada, to northern Mexico, and eastward from Denver to about the 100th longitudinal meridian, which has been the traditional dividing line between the deep-rooted tall grasses with large leaves of the tall-grass prairie and the shallow-rooted, short grasses with small leaves.

Along the trail of Lewis and Clark, the 100th longitudinal meridian passes north-south near the mouth of the Niobrara River, that is to say, near where the Missouri River arrives at the Nebraska border and begins to form the Nebraska–South Dakota line. This was near where Lewis and Clark first saw prairie dogs—a location within fifteen miles of modern Fort Randall Dam. Finding prairie dogs reinforced the fact that Lewis and Clark had crossed another major boundary between ecological regions—from the tall- and mid-grass prairie to the short-grass prairie, with the Niobrara as the major eastern boundary of the short-grass prairie region.

Prairie Dog Country

Lewis and Clark saw prairie dogs and the short-grass prairie across a long distance—from the Niobrara River to west of Bismarck, North Dakota. At the time of Lewis and Clark, the black prairie dog (*Cynomys ludovicianus*) was found from Saskatchewan to northern Mexico. There are four other species of prairie dogs, all native to North America. Prairie dogs are members of the squirrel family and closely related to other ground squirrels common elsewhere in North America. Today an estimated 98 percent of prairie dogs in the United States have been eliminated, most by poisoning, because they have been thought to compete with cattle for grasses and because their holes pose a problem to horses and cattle that injure their legs

Fort Randall Dam. The Fort Randall Dam at Pickstown, South Dakota, was built in 1954 by the Army Corps of Engineers. It is near where Lewis and Clark first saw prairie dogs. What was then a vast short-grass prairie with an extensive, underground reserve of water is now farmland surrounding a large reservoir on the Missouri River. The dam produces enough hydroelectric energy for 245,000 homes. *U.S. Army Corps of Engineers.*

when they step into them. Prairie dogs also suffered loss of habitat as the short-grass prairie was converted to cropland and grazing land. The prairie dog is a controversial animal. Some have thought of it as an enemy of the farmer and rancher or as good rifle target practice; others believe it is an adorable creature that should be protected.

Two of its predators are listed as rare or endangered and are protected by state and federal laws—the black-footed ferret and the burrowing owl. Conflicting views about prairie dogs can be found on the Web, even between websites from the same state. Not long ago—in 2000—one website for the State of South Dakota has stated that "the black-tailed prairie dog is found throughout western South Dakota. It's no secret these small, gregarious rodents are a major irritant to stockmen whose cattle compete with the burrowing grass-eaters for grazing land. There is no closed season, so it's legal to shoot prairie dogs anytime. The ideal time is from May through September." According to that website, any caliber rifle or handgun is legal, and there is no limit on the number of prairie dogs an individual may shoot.

At the same time, another state of South Dakota website explained that "The State of South Dakota has been actively involved in prairie dog management planning since 1999. The goal of the state's planning effort is to manage for long-term, self-sustaining prairie dog populations in South Dakota while avoiding negative impacts to landowners that do not wish to accommodate prairie dogs on their properties."

The black-footed ferret, a predator of the prairie dog, has been called the most endangered mammal in North America. According to the American Zoo and Aquarium Association, "the decline of the black-footed ferret is almost entirely due to government-sponsored poisoning of prairie dog towns and development of farms, roads, towns, etc. over prairie dog colonies." Prairie dogs make up about 90 percent of the diet of the ferrets, which inhabit abandoned prairie dog burrows, so they depend on the prairie dogs for food and shelter. According to this conservation association, "recent studies have proven that the grass-eating prairie dogs are not significant competition with livestock for forage," and the problem is one of education rather than of competition.

On the one hand, money is made available to promote hunting of the prairie dogs, which in turn threatens prairie dog habitat and therefore the habitat of the black-footed ferret. On the other hand, money is made available for captive breeding of the ferret with a plan to reintroduce the fer-ret—whose success depends on the abundance of prairie dogs—into its original habitat. So the state government and the ranchers argue that prairie dogs must be reduced because they compete with cattle; conserva-tionists argue that prairie dogs do not compete with cattle; and the U.S. Fish and Wildlife Service seeks to protect the ferret because it is listed as endangered under the U.S. Endangered Species Act. So it is with many environmental issues.

However many champions or enemies the prairie dog may have, the black-footed ferret has been of great concern to conservationists. The ferret population had been reduced to very low numbers by the 1950s. In the 1960s, only one population was known, a small number of ferrets in South Dakota. But the colony's last member died in 1979, and the species was then believed to be extinct. Then a population was discovered in Wyoming, and the state's Game and Fish Research Facility began a captive breeding pro-gram with eighteen ferrets. The population has grown to more than three hundred, some of which are now in zoos around the nation and some of which have been reintroduced near prairie dog towns in the United States and Canada.

The problem here is about ecological food webs—the relationships between predator and prey. The most endangered species of the short-grass prairie, the black-footed ferret, survives near the top of the food web, as a predator of the prairie dog. The prairie dog eats vegetation for the most part, except for an occasional protein-rich grasshopper. In general, the far-ther up a food web, the less abundant a species and the more likely the species is to become threatened and endangered. The ferret's problem is being at the top of the food chain, a situation shared by many endangered and threatened species.

Viewing Prairie Dogs Today

Plowing, grazing, and intentional killing have removed prairie dogs from most of their range. What was once a common sight in the western plains is now a rarity, and watching the behavior of these animals in their villages is worth taking the time to enjoy. One place to see prairie dogs today is the Little Missouri National Grasslands, which extends north from U.S. 12 in

Bowman County, North Dakota, and lies east of the north-running Little Missouri River.

According to the *North Dakota State Parks and Recreation Outdoor Adventure Guide*, this is the largest "and most diverse" of the nineteen grasslands found in the western United States. It contains rolling prairie, badlands, and woodlands in the valleys. The Little Missouri River is North Dakota's only designated scenic river. It flows south to north through the park and the national grasslands, and extends for a total of 274 miles until it enters the Missouri River. The Little Missouri has carved a kind of badlands through this dry country.

A road takes a visitor along a thirty-six-mile scenic loop, on a two-lane paved road that climbs into badland-like bluffs eroded by the Little Missouri River. There are several waysides near prairie dog towns, including some picnic grounds. There you can see prairie dogs doing what Clark described them doing—sitting erect, whistling, appearing out of their holes to observe their surroundings, and foraging for food.

Unlike Lewis and Clark, a visitor today should think twice about handling prairie dogs because they carry a form of plague spread by fleas; if a prairie dog is touched by a person, the plague can spread to human beings. You can be bitten by one of these fleas even when you dig around in their burrows.

Surviving a Cold Winter

Lewis and Clark brought a thermometer along with them and recorded the temperature at noon and at four in the afternoon whenever possible. During their winter with the Mandans, they recorded temperatures in January 1805 that ranged within a few weeks time from -40 to +35 degrees Fahrenheit. The next summer, Clark and Sacagawea were almost drowned by an intense thunder- and hailstorm; within a month, Clark would walk in dry heat that gave him sunstroke. Extremes of weather seemed to be the rule. Part of the reason for these extremes is that the Missouri River basin lies totally within the interior of a major continent far from the ameliorating effects of an ocean.

Interiors of continents are notorious for their highly variable climates. Fast changes in temperature, like those Lewis and Clark experienced in January 1805 in the Missouri River Basin, occur at the crossroads of major air masses. Some come from the north, some from the south, generally one alternating with the other. This leads to large changes in temperature over a few days, even within a few hours. In the basin as a whole, temperature extremes range from -20 degrees Fahrenheit in the winter to 110 degrees in the summer, although Lewis and Clark measured even deeper cold.

On October 18, 1804, there was a hard frost, freezing clay near the river as well as water in containers. The day after they arrived at the Hidatsa and Mandan villages, a half inch of snow fell. A week later, all the leaves on the trees had fallen, including those of the cottonwoods. Snow fell but did not stay on the ground. Violent winds struck on October 28 and 29, and in this way winter came. The expedition would stay here until April 7, 1805—more than five months. They began to build their winter camp on the second of November, calling it Fort Mandan.

Lewis and Clark had the men build huts of cottonwood, "this being the only timber we have," Clark wrote on November 6, 1804. On November 3, in his journal, Sargent Gass described the huts as being "in two rows, containing four rooms each, and joined at one end forming an angle." The floor was of split planks, covered with clay and grass. The roofs reached about eighteen feet above the ground. Two storerooms were built in the angle formed by the other huts. During the building of the fort, two men cut themselves with axes. Meanwhile, ducks, geese, and brants continued to pass overhead. On November 13, the men moved into their huts and began their stay for the winter. The next day, Clark wrote that the ice was "runing verry thick" as the river rose and some snow fell. On the sixteenth, "a verry white frost" covered "all the trees" with ice, and all the men moved into huts, finished or not.

Along with, and as a consequence of, the varying temperatures, winds vary greatly in the Missouri River basin. They can reach one hundred miles per hour. As a rule, winds are strongest in the spring. The high winds evaporate water and dry the soil faster than still air. Blizzards—snow storms with strong winds—are a common winter hazard.

With the huts finished enough to protect them from the worst of the weather, the men of the expedition turned to hunting—out of necessity. On November 19, the hunters returned with "32 Deerr, 12 elk & a Buffalow," the first mention in the journals of buffalo since late October, most likely because Lewis and Clark had been occupied with preparation for the winter rather than hunting. The weather improved toward the end of November. November 23 was a "fair warm Day," as were the next two, but then the weather turned again and became very cold and windy.

On the thirtieth, the Mandan chief told Lewis and Clark that some of his hunters had been attacked and killed by Sioux. Clark wrote that "we thought it well to Show a Disposition to ade and assist them against their enimies." Lewis and Clark took twenty-three men armed and on horseback with a promise to defend the Mandans, but the chief suggested that they wait until spring because the snow was too deep for the horses and it was too cold.

Winter was closing in upon them, their crude huts were barely completed, they were not well stocked with food, but they took up arms and went out with their horses to help their hosts against the enemies. So began their first winter on the plains.

On December 7, 1804, the chief of the Mandans told Lewis and Clark that there were "great numbers of Buffalow" on the hills nearby. Lewis went with a hunting party of fifteen of his men and killed eleven, "three in view of our fort," but the weather was "so excesive Cold & wolves plenty, we only saved 5 of them." Clark went with fifteen men and four horses the next day, when the temperature was 44 degrees below freezing, and found buffalo about seven miles away. They killed eight, but several of the men suffered frostbite. On the ninth, Lewis went out and stayed out all night, experiencing "a Cold Disagreeable night . . . in the Snow on a Cold point with one Small Blankett." The ice was so hard on the river that the buffalo crossed without breaking through. Lewis and his men killed nine buffalo, but many were "So meager that they [were] not fit for use." On the twelfth, pronghorn were seen, but Clark wrote that the weather was "So Cold that we do not think it prudent to turn out to hunt." Clark, however, with his proclivity for making measurements, paced the width of the river by walking across on the ice: He found it to be five hundred yards wide.

Clark went on a hunting party on December 7 and 8, when there was much snow and it was 44 degrees below freezing. They continued to see buffalo from December 14 to 18, but again the weather was too cold for hunting. Their journals suggest that they were visited by a large herd, which remained in the vicinity of the camp for ten days. They next reported buffalo on January 6, 1805 when Clark was out hunting them with sixteen men, killing a total of eight.

On January 14, Clark wrote that one of their hunters, who had been sent out for several days, returned to say that another member of the expedition, Whitehouse, was so badly frostbitten that he could not walk home. In spite of the cold, Lewis reported on the same day that there was an eclipse of the moon, which he observed with the small refracting telescope that was part of his sextant, having "no other glass to assist me in this observation." He was able to "define the edge of the moon's immage." He wrote that clouds interrupted his observations, which made the observation of the "*commencement of total darkness*" inaccurate. "The two last observations (i.e.) the *end of total darkness* and the *end of the eclips*, were more satisfactory," he wrote. These observations Lewis used to locate the longitude of their winter fort. Thus he continued, in spite of the cold, to map carefully the expedition's course across the continent—to make quantitative measurements, in keeping with the scientific purpose of the expedition.

We often think of "man and nature" as a primitive relationship—men pitting themselves against nature without the aid of civilization. In fact, technologies played a key role in the success of the expedition, as Lewis's use of his sextant to observe the moon testifies. Most important to the expedition along with the sextant were the compass; the gun; blacksmithing tools; the knowledge of surveying and making wheels, wagons, and ax handles; and, of course, writing. At times, the blacksmith of the expedition traded his skills for Mandan corn and repaired many objects. The gun saved several of the men from grizzlies later in the trip, and hunting with guns provided the staple food: meat.

On February 4, Lewis wrote that "no buffaloe have made their appearance in our neighbourhood for some weeks" and that the "stock of meat" was "nearly exhausted." Clark decided to take a group of men down the river to hunt. They pulled their baggage on small wooden sleighs and brought three packhorses to carry the meat they acquired. On their first day, they found nothing to hunt and had nothing to eat. On the second, Clark broke through the ice, and his feet and legs got wet. That same day, they killed a deer and two buffalo, but the buffalo were in too poor condition to eat. They saw buffalo on February 8 but found them again too lean to be worth taking. Clark returned on the night of the twelfth. He and his men had walked thirty miles on the ice and through the woodlands. In some places the snow was up to their knees. On this trip, they killed a total of forty deer, three buffalo, and sixteen elk. They saw no more buffalo that winter.

8

AMERICA'S SERENGETI

I F ANY LOCATION on Lewis and Clark's journey matched the idealized view of the presettlement American West as a Garden of Eden rich in vegetation with great numbers of wildlife, it was the confluence of the Yellowstone and Missouri rivers, just east of today's Montana–North Dakota border. The expedition arrived there on April 25, 1805. Lewis took four men and went by foot to explore the Yellowstone River upstream from its mouth. From a hilltop, Lewis wrote, "I had a most pleasing view of the country, perticularly of the wide and fertile vallies formed by the missouri and the yellowstone rivers, which occasionally unmasked by the wood on their borders disclose their meandering for many miles in their passage through these delightfull tracts of country."

He continued that "the whol face of the country was covered with herds of Buffaloe, Elk & Antelopes; deer are also abundant." Adding to this image of Garden of Eden, he wrote that "the buffaloe Elk and Antelope are so gentle that we pass near them while feeding, without appearing to excite any alarm among them, and when we attract their attention, they frequently approach us more nearly to discover what we are, and in some instances pursue us a considerable distance apparently with that view."

The vegetation was also abundant. Lewis wrote the next day that "there is more timber in the neighbourhood of the junction of these rivers . . . than there is on any part of the Missouri above the entrance of the Chyenne river to this place." On the floodplains were cottonwood, "small elm, ash and boxalder" along with "Goosbury, choke cherry, purple currant; and honeysuckle bushis." On sandbars in the river were willows, wild roses, and serviceberry. Where there were no trees, there were many small plants, including "wild hyssop which rises to the hight of two feet" and which was a favorite food of "the Antelope, Buffaloe Elk and deer." Willows filled river sandbars and "furnish a favorite winter food to these anamals as well as the growse, the porcupine, hare and rabbit," he added.

Happy to Be There

It was an American Serengeti, and everybody was happy to be there. Lewis wrote that "to add in some measure to the general pleasure which seemed

The Confluence of the Yellowstone and Missouri Rivers. Although still attractive, the confluence lacks the great abundance and diversity of wildlife that greeted Lewis and Clark. A rural setting, the confluence has boat ramps and is frequented by recreational fishermen. *D. B. Botkin.*

to pervade our little community, we ordered a dram to be issued to each person, this soon produced the fiddle, and they spent the evening with much hilarity, singing & dancing, and seemed as perfectly to forget their past toils, as they appeared regardless of those to come."

The physical—geological—scene is similar today, and the confluence is known for its abundance of pronghorn, deer, and migrating birds. The animals are likely less tame, and the elk and bison are gone, so the overall biological diversity has probably decreased (although to my knowledge no one has measured it).

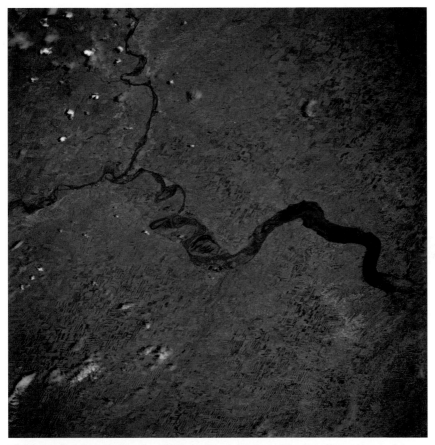

Missouri–Yellowstone Confluence, visible from space in this image taken in April 1994. The two rivers come together in the left center. To the right is the western part of the western end of Lake Sakakawea, the reservoir formed by Garrison Dam on the Missouri River. *Image courtesy of Earth Sciences and Image Analysis Laboratory, NASA Johnson Space Center. STS059-89-31, http://eol.jsc.nasa.gov*

Lewis and Clark as Scientists

Lewis and Clark saw many species and provided the first written description of three hundred of them—178 plants and 122 animals. Lewis's collection of dried plants is maintained in the herbarium collection at the Philadelphia Academy of Sciences. These specimens were used by the German botanist Frederick Pursh, who included Lewis's plant discoveries in *The Flora of North America*, published in 1813. Some, like Snow-on-the-Mountain (*Euphobia marginata*), collected by Lewis in 1806, became common in seed catalogues in the nineteenth century and helped decorate many a garden. Lewis's plant collections continue to be useful today.

While these three hundred descriptions were and are useful to science, arguably more important is the way the journals illustrate Lewis and Clark's greatness as naturalists, observers, and interpreters. Today in our scientific age it might be easy to criticize Lewis and Clark for being "amateur" scientists, but this does not make sense within their time. There were no professional scientists in North America at the beginning of the nineteenth century, if we take "professional" in the modern sense of being paid to do something and that something is one's primary way of making a living.

At the beginning of the nineteenth century, botany, zoology, and geology were considered part of "natural philosophy," and any interested person was welcome to take part in them. Jefferson was a great natural philosopher. So too were many physicians of that time, perhaps because they tended to have microscopes, and perhaps because they would have pursued medicine through a fundamental interest in what we now call science.

This criticism of Lewis and Clark as naturalists and scientists also assumes that, lacking the best, most extensive formal training, they could not have been good observers. But this is contrary to the facts. One story illustrates this. About twenty years after Lewis and Clark's voyage, a group of Europeans came to the Missouri River as part of a "scientific" expedition. The scientists in this group caught a catfish from the Missouri River. Because there was only one species of catfish in Europe, these scientists assumed that it was the same species and did not provide a careful description of the fish, referring to it only as "the catfish." There were, however, five species of catfish in the Missouri River. As a result, this record by the "professional scientists" is useless today. In contrast, Lewis, not knowing how many catfish species there might be—and, more important, not making any presumptions about that number—acted as an accurate observer and recorder. His journal provides information sufficient to determine the exact species he saw and is valuable to fisheries scientists today.

As the story was told earlier, Lewis and Clark began to see animals of the short-grass prairie the previous fall, in 1804, when they reached the 100th meridian and were closing in on the Mandan villages where they would spend the winter. On September 5, Clark recorded the expedition's first encounter with mule deer and pronghorn, providing the first written scientific record of the mule deer, which they named. Clark also mentioned "Several wild goats on the Clift & Deer with black tales." Entering the short-grass prairie the expedition had entered the landscape and wildlife that we identify today as the American West—cowboy-and-Indian country of open plains and big skies. They began to encounter a wealth of new species.

On September 16, they saw and shot a black-billed magpie, about which Lewis wrote a characteristically detailed description. He noted that "the wings have nineteen feathers, of which the ten first have the longer side of their plumage white in the midde of the feather" and the "upper side of the wing, as well as the short side of the plumage" was "a dark blackis or bluish green sonetimes presenting as light orange yellow or bluish tint as it happens to be presented to different exposures of ligt." He wrote that "it is a most beatifull bird," with the outer wings changing color in different lights to be an orange-green and then a reddish indigo blue. He was observing one of the many kinds of birds whose plumage color changes with the refraction of the light.

On September 17, when the expedition neared the site of modern-day Chamberlain, Clark wrote a longer description of mule deer when one of the men, Coulter, shot one. Clark observed that it was "a Curious kind of Deer, a Darker grey than Common the hair longer & finer, the ears verry large & long a Small resepitical under its eye its tail round and white to near the end which is black & like a Cow," but "in every other respect like a Deer, except it runs like a goat," later adding that it "jumps like a goat or Sheep" and that it was "large." On that day, one of the hunters also brought in "a Small wolf with a large bushey tail," probably the expedition's first coyote. The description was another of their contributions to biology.

The mule deer is the characteristic deer of the western plains and Rocky Mountains, living in conifer forests, desert shrublands, and prairies. At the time of Lewis and Clark, the mule deer spanned the territory of the Dakotas west to the Pacific coast. Along the coast, they were found from Baja California to British Columbia. In the prairie, they ranged from the southern edge of Alaska through Canada and south into central Texas and interior Mexico. Mule deer feed primarily on herbaceous plants, including grasses, but also browse on shrubs and trees. The whitetail deer, the characteristic deer of the eastern United States, is more of a forest animal, primarily feeding on leaves, twigs, fruit and nuts of shrubs and trees. At the time of the expedition, the whitetail had a larger range: it was found in Lewis and Clark's home state of Virginia, as well as across the entire eastern seaboard. It was absent at that time only from the Rocky Mountains and California. Both deer remain abundant.

The pronghorn antelope is native only to North America. It has been here for about fifty million years, since the Eocene epoch of geology. Pronghorn feed on grasses and forbs during the growing season, and eat shrubs in the fall and winter. Some experts have estimated that at the time of the Lewis and Clark expedition, pronghorn may have been almost as abundant as buffalo, perhaps numbering thirty or forty million. But soon after that time, pronghorn became a major item in the diet of pioneers, and it was also sold commercially. A creature of the prairie, the pronghorn lost its habitat to the plow and cattle. By the beginning of the twentieth century, there were only about ten thousand pronghorn remaining.

Concern about the decline of this species occurred rather early in the history of conservation in America. In the early twentieth century, states began to pass laws protecting the pronghorn and outlawing hunting. Yearly counts of the herds began, as well as collection of information about diseases and predation. In the 1920s, the first extensive census was made over the entire range, and there appeared to be about thirty thousand pronghorn. Today, these animals number over a million, with the greatest numbers in Montana, Wyoming, and North Dakota. They remain relatively rare in South Dakota—under ten thousand. The pronghorn population has grown for the past eighty-five years, and it more than doubled between the late 1970s and late 1980s. With care and restoration of prairie habitat, the outlook for the pronghorn remains good: it is now possible to see pronghorn along Interstate 90 east and west of Gillette, Wyoming, and on the back roads of central and eastern Montana, although most likely you will still have to seek wildlife refuges and prairie preserves to see these animals.

Even more useful to us than the descriptions of individual species are Lewis and Clark's experiences with them—especially the larger, more spectacular mammals of the American West. Their journals are a rich source of information about wolves, grizzlies, and bison in presettlement America. The rest of this chapter is devoted to these animals.

Wolves and the Rationales for Saving Endangered Species

The expedition encountered wolves at least as early as June 30, 1804, at the mouth of the Platte River in Nebraska. On that day, Clark wrote that they "set out verry early this morning," when "a verry large wolf Came to the bank and looked at us." He paid no more attention to it in his notes, however—wolves, in spite of their reputation, are not dangerous to human beings, and Clark would have known this. Still, the look in the eyes of a wolf—one of the great images of the North American wilderness—would have been impressive to most viewers. In my own experience, the steely, pale eyes of a wolf and the cold blue eyes of an African leopard are views into a kind of wilderness hard to express. A week later, on July 7, Lewis and Coulter shot and killed a wolf.

Although they had seen wolves during the journey, this species was especially common below and above the confluence of the Yellowstone. On April 29, 1805, near the mouth of Martha's River, they found themselves "surrounded with deer, elk, bison, antelopes, and their companions the wolves, which have become more numerous and make great ravages among them." A week later, on May 5, Lewis wrote about "Buffaloe Elk and goats or Antelopes feeding in every direction" and "a great number of" wolves.

The "constant companions" of the bison and other large herbivores, as Lewis and Clark put it, are making a comeback, after almost being extinguished in the lower forty-eight states and along the Lewis and Clark trail. Most of the recovery habitats, however, are not along the path that Lewis and Clark followed.

Today, for some people, the presence of wolves would be the final touch in creating a wilderness paradise at the confluence of the Yellowstone and Missouri. For others, the presence of wolves would destroy the very idea of a Garden of Eden. People have hated wolves throughout most of western history, and the desire to conserve wolves is relatively new to western civi-

The Gray Wolf, Today. The "constant companions" of the bison and other large herbivores as Lewis and Clark put it, are making a comeback, after almost being extinguished in the lower 48 states and along the Lewis and Clark trail. Most of the recovery habitats, however, are not along the path that Lewis and Clark followed. *U.S. Fish and Wildlife Service/photo by Gary Kramer.*

lization. Ancient Greek and Roman writers, including Aristotle and Plutarch, mention the evil and dangerous nature of the wolf. In Dante's *Inferno*, the wolf represents human greed. In contrast, some American Indian tribes had wolf clans, considering the wolf to be a fetish and a "brother."

Today the wolf represents a powerful symbol of wild nature. In its wariness of people, the wolf epitomizes our predominant contemporary image of nature: nature separate from human beings and human beings divorced from nature. Where we are, there are no wolves; where the wolf lives, there is wilderness.

Because of their contradictory symbolism, wolves—perhaps more than any other large animal of the American West—make us ask why we should save endangered species and why we should expend large sums and restrict land use, hunting, fishing, and other activities on their behalf. The great diversity and abundance of life at the confluence of the Yellowstone and Missouri rivers makes the question all the more compelling. The controversy has focused upstream, in Yellowstone National Park, where wolves have been reintroduced in areas where they had been locally extinct for sixty years.

One standard answer is utilitarian: a species directly benefits people or may in the future. One of the purposes of the expedition was to find such benefits, including determining the potential for fur trade with the Indians of the West. Lewis and Clark therefore carefully recorded the distribution and abundance of beaver, as well as beaver lodges, dams, and tree cutting. Even grizzlies, which were dangerous to the expedition, had a purpose—their hides were used by the expedition. In contrast, wolves—another other big mammalian predator the expedition saw—were neither of use nor a threat. If there has ever been a reason to conserve wolves, it would seem to lie beyond a direct, practical benefit to people.

The second argument for biological conservation is that a species plays an essential role in its ecosystem. It is an ancient question, found in Greek and Roman writings: Why should there be predators—considered vile and vicious creatures—on the Earth? The answer has always been that these animals control the abundances of their prey. This argument was picked up by modern ecology and formulated in terms of mathematics and mechanical physics that conceived wolves, in theory at least, as mechanical devices, personality-less entities that bang into their prey at random.

That theory, still prevalent, predicts that such predators and their prey will function together to control each other's numbers with great precision and predictability. According to this theory, the prey species would increase uncontrollably without its predator. The prey would overeat its food supply, and its population would crash, perhaps leading to its extinction. But with both predator and prey present, the two would either achieve a constant abundance that would persist indefinitely, or they would oscillate forever, exactly out of phase, like two guitar strings tuned to the same note, the second plucked at exactly the moment when the first reached its peak of vibration. The only problem with this theory is that it fails to predict anything real. All field studies show that real predators and prey do not follow these predictions. Big-game predators can reduce the numbers of their prey, but they have never been observed to control the abundance in the precise way the theory predicts.

This theory arises from the same balance-of nature worldview that led us to believe we could channelize the Missouri River, run it as if it were a hydrological machine, and receive its benefits with no ill effects. It is the same worldview that ignores the connection between human settlements and rivers and that has led us to build big interstates between cities and their riverfronts. It has failed with predators just as it has failed with rivers and cities.

Another standard justifications for the conservation of endangered species is moral. Wolves are social animals, species with highly developed social behaviors and signs of individualism that appeal strongly to many people. Such social attributes have also been used to support conservation of whales and porpoises, whose care for their young and apparent intelligence leads many to propose moral arguments that these creatures have a right to exist. Wolves share these qualities. They typically live in packs of four or five to twenty. There is a rigorous social structure, with a lead male and female who breed and whose pups are cared for by other adults as well as by the parents. The lead male affirms his dominance through his posture and, when challenged, in fights. The personality of the lead male seems to influence the behavior of the entire pack. He does not happen upon his prey at random.

But in my opinion still another justification provides the underlying rationale for most people to save wolves. This justification is aesthetic. I have worked in wilderness areas with wolves. What I remember most about them is their calls at dusk and during the night. Few other sounds bring out the wildness of the woods, evoking the essence of wilderness and the connection between human beings and nature. For me, these feelings are powerful justifications for saving the wolves.

Nevertheless, people shy away from the aesthetic argument like a wolf shying away from people, as if nobody would take it seriously. Most discussions instead emphasize the utilitarian and the ecological, both mechanistic approaches. They are echoes of the machine age of the nineteenth and early twentieth centuries, when the science of ecology first attempted to explain how nature works. To so limit the value of big predators is to debase the deep connection between the human spirit and nature. We have passed through an era of arguments limited to the utilitarian, and we have seen the consequences of it in our rivers, our cities, and here, in the diversity of life.

New developments in ecological science point to ways we can deal with the complexities of real predators, with their individual behaviors, social interactions, and effects on their environment. And so the question before us at

Some Wildlife, Seen by Lewis and Clark, Such as Pelicans, Are Sill Abundant. *U.S. Fish and Wildlife Service/photo by Rod Krey.*

the confluence of the Yellowstone and the Missouri rivers is whether there is a place—in our imaginations and in our realities—for wolves in the American West. Do we want to envision the richness of life with wolves or without them? With all the intricacies and complexities of nature, or without them? When you visit the confluence and see the still well-watered bottomlands and the Yellowstone River, neither channelized nor dammed and therefore more in its presettlement condition than the Missouri, it is a time to reflect on this question.

Grizzlies: Estimating Original Abundance

Although wolves have had a negative image in Western civilization, they were not dangerous to Lewis and Clark—in contrast to the grizzlies, the most dangerous mammals the expedition encountered. On May 11, 1805, when the expedition was northeast of what is now the Pine Recreation Area near Fort Peck Dam in eastern Montana, Bratton, one of the members of the expedition, went for a walk along the shore. Soon after, he rushed up to

Lewis "so much out of breath that it was several minutes before he could tell what had happened." Bratton had met and shot a grizzly bear, he told Lewis, but the bear didn't fall: instead, it ran after him for about half a mile, and it was still alive.

Lewis took seven men and trailed the bear about a mile by following its blood in the shrubs and willows near the shore. Finding the bear, they killed it with two shots through the skull. Upon cutting it open, they found that Bratton had shot the bear in the lungs, after which the bear had chased him and then moved in another direction, a total of a mile and a half. "These bear being so hard to die reather intimedates us all," Lewis wrote. "The wonderful power of life which these animals possess," the journals continue, "renders them dreadful; their very track in the mud or sand, which we have sometimes found 11 inches long and 7 1/4 wide, exclusive of the talons, is alarming."

This was not their first encounter with a grizzly—that had taken place the previous fall, on October 20, 1804 when they were near Bismarck, about to set up their winter camp. That location, in the Great Plains hundreds of miles east of the Rocky Mountains, considerably extends the eastern-range boundary assumed for this animal. Lewis and Clark saw grizzlies during the next spring and into the summer. On approximately twenty days between April 17 and the end of July, they saw these bears—about one encounter or sighting every five days. Most of their sightings were upstream from the confluence of the Yellowstone and Missouri. They were especially troubled by them when they were portaging their equipment around the Great Falls. Their last sighting was near Three Forks, Montana, the headwaters of the Missouri. No grizzlies were found east of Pierre, South Dakota, nor west of a north-south line passing through Missoula, Montana: the grizzlies were confined to two regions of the trip—the upper Missouri and adjacent short-grass prairies, and the foothills of the Rocky Mountain forests—the dry plains and the cold mountains.

Also in the spring of 1805, not far from where Bratton was chased by the grizzly, Lewis wrote the first scientific description of this species, although it did not receive its scientific name, *Ursus horribilis*, until 1815. Lewis described a male "not fully grown" that he estimated weighed three hundred pounds and that they had killed after shooting it many times. He wrote that the grizzly had longer legs than the black bear, that its color was "yellowish brown, the eyes small, black, and piercing." "[T]he front of the

Hunting Grizzlies. Karl Bodmer's impression of grizzlies, as fierce creatures, and as he saw them along the Missouri River not long after Lewis and Clark passed that way. Karl Bodmer, *Hunting of the Grizzly Bear,* hand-colored engraving (Tableau 36). *Joslyn Art Museum, Omaha, Nebraska.*

fore legs near the feet is usually black; the fur is finer thicker and deeper than that of the black bear," Lewis reported.

Because grizzlies are so big and dangerous, Lewis and Clark recorded the number of bears (usually one) in each encounter. Reading their accounts, I realized that it was possible to use the journals to estimate the original abundance of these dangerous animals and to learn about their original range. The expedition encountered a total of thirty-seven grizzlies over a distance of approximately one thousand miles, an average of about four grizzlies per one hundred miles traveled. The area known to have grizzlies today, twenty thousand square miles, is 6 percent of the presettlement range of the bear, based on the journals of Lewis and Clark. Today, grizzly habitat exists mainly on government land, mostly U.S. Forest Service land, in four states. Only 5 percent is private land. Much of the rest is in four national parks: Glacier, Yellowstone, Grand Teton, and North Cascades.

Habitat in and around Yellowstone National Park that appears to have grizzlies is estimated currently at about 7,800 square miles. You are very unlikely to see a grizzly, but at the Pines Recreation Area and elsewhere, you can see grizzly bear habitat. The rare encounter with a grizzly today could occur if you went cross-country backpacking in one of the national parks or national forests. You are more likely to see them in the Canadian Rockies, although there, too, the chances are low: the chances are greater in Alaska.

Why would anyone want to know how many grizzlies there were? Grizzlies are listed as an endangered species, and the U.S. Fish and Wildlife Service has a recovery plan for the grizzly bear. But recovery to what? Under current interpretation of the Endangered Species Act, a species can be listed as threatened or endangered if its numbers drop to less than one-half of the estimated "carrying capacity"—the maximum number of animals that a habitat can support. And the carrying capacity is typically taken to be the estimate of presettlement abundance. That number can be estimated from Lewis and Clark's journals.

The density of the bears was about four for every one hundred square miles. We assume that on average the men of the expedition could see a half-mile on each side of the river. Using this average along with the assumed presettlement range of the bears, about 530,000 square miles, we estimate that there might have been as many as twenty thousand bears.

The Grizzly Bear. Lewis and Clark considered the grizzly the most dangerous animal they encountered. Its size and ferocity make it a symbol of the great American wilderness—but almost led to its being hunted to extinction. It is now listed as an endangered species; the grizzly is protected by the Federal Endangered Species Act. Here a grizzly feeds on a dead elk in Yellowstone National Park, where one of the largest populations of these animals live today. *U.S. Fish and Wildlife Service/photo by LuRay Parker.*

Recently, two scientists made use of the method that I first suggested to develop another estimate of early nineteenth-century grizzly abundance from the Lewis and Clark journals. These scientists, Andrea S. Laliberte and William J. Ripple, expanded on the original idea. They reviewed the journals, using both the westbound and the eastbound journey. I avoided the eastbound journey because I assumed that the members of the expedition were hurrying home and not as likely to observe wildlife. However, the comparison is interesting.

Laliberte and Ripple determined that the expedition killed a total of forty-three grizzlies. Lewis and Clark reported seeing forty-two grizzlies, while twice reporting "many" grizzlies and four times reporting "some." I had ignored the some and the many, but they interpreted some and many to be two grizzlies each. From these records, they estimated that at least ninety-seven grizzly bears were encountered, or about forty-eight each way. That would yield about five for every one hundred square miles and a total of 20,000 to 27,000. Such estimates are quite approximate, and the entire exercise very rough, but it is the best we can do.

Although we are legally required to restore the grizzlies, and an estimate of presettlement abundance is the usual method, I was surprised to find that there are few other studies that provide any useful estimate of their abundance. One of these was made by the Craighead brothers, two of America's experts on grizzly bears. Their study was limited to Yellowstone National Park, where they reported an average of 230 grizzlies between 1959 and 1967, an average density of three bears per one hundred square miles, similar to my estimate from the journals.

Strangely, with the sole exception of information gathered in Yellowstone, our present knowledge of the abundance and density of grizzlies is not much better than what someone could have surmised from Lewis and Clark's journals in 1806. If this is what we know about one of the most famous, readily reported, legally threatened, and therefore protected species, whose abundance and whereabouts are of considerable interest to outdoorsmen and government agencies, what could be our knowledge of other species? The answer is, in most cases, much less.

But is the goal of restoring populations to their presettlement size the right one? This approach assumes the constancy of nature—that before the influence of European civilization, the abundance of grizzlies and everything else in nature never changed from year to year. This doesn't make much sense, and all the evidence available about wildlife suggests that it has never been true: populations of wildlife change all the time. Such a belief, while consistent with the ancient idea that nature is balanced, contradicts the changeableness of the environment, which Lewis and Clark came to know all too well in their travels on the Missouri.

Scientists now know that populations of grizzlies and other animals and plants are, like the Missouri, always changing. There is no single "natural" abundance. There is a range of abundances, all of which are "natural" in the sense that populations expanded and contracted prior to effects of modern civilization. This has become known as the "historic range of variation."

When we recognize this, a plan to return the grizzlies to their "original" abundance becomes more complicated. We begin to wonder not what the right number is, but what are the keys to sustainability.

More recent programs to restore endangered or threatened animals have begun to focus on this more realistic goal of a self-sustaining population. Apparently, this was the goal for a Fish and Wildlife Service's Grizzly Recovery Plan. Its objective was "to establish viable, self-sustaining populations in areas where the grizzly bear occurred in 1975." To realize that goal, we must understand much more about the requirements of this species. We must understand what it needs from its habitat and the ecosystems within which it lives. We have to obtain estimates of the abundances of the bears before and after settlement by Europeans and, if possible, obtain estimates from different times so we can calculate the range of variation.

On June 28, 1805, the expedition was in the midst of portaging around the Great Falls of Montana. Lewis noted in his journal, "The White bear have become so troublesome to us that I do not think it prudent to send one man alone on an errand of any kind, particularly where he has to pass through the brush." The bears were bold enough to "come close arround our camp every night but have never yet ventured to attack us and our dog gives us timely notice of their visits, he keeps constantly padroling all night." It was so dangerous, Lewis believed, that "I have made the men sleep with their arms by them."

In summary, to believe that there is a single magic number that is the only sustainable one is to believe that a species is fragile and that individuals within a population are not resourceful. This seemed hardly the case with the grizzlies that met Lewis and Clark. The grizzlies were fearless, strong, able to withstand a number of bullet wounds; they were quick to respond, resourceful. A population that persists and prevails over a long time must have abilities to respond to change. To be sustainable is different than to persist for a time at a certain level of population, and that to maintain a steady population may not be the best strategy for a species.

Lewis and Clark's encounters with grizzly bears were their most dangerous encounters with animals and among the most dangerous of all their experiences. But the meaning of these encounters in our search to come to know nature in the American west is much greater, much deeper. From their encounters with the grizzlies, we learn much. We learn about the limits of our present knowledge. We learn that, in spite of much emotion and desire directed toward the conservation of rare and endangered animals during the past thirty years, our knowledge remains terribly limited. We discover that we know little more about the range and density of the grizzly bears in the lower forty-eight states than one would have known from reading Lewis and Clark's journals in the early nineteenth century. We discover that clear, objective, written historical records can be of great help to us. And in the end, we discover that we have a much longer journey ahead of us than Lewis and Clark if we are to be able to succeed our attempts to conserve endangered species.

Bison—Near Extinction in a Few Decades

On May 29, 1805, the expedition reached the mouth of the Judith River where that river enters the Missouri in Montana. Clark named the river for Judith Hancock of Fincastle, Virginia, whom he would marry in 1808. On that day, Lewis wrote that near to this junction they passed the "remains of a vast many mangled carcases of Buffalow." He believed that Indians had driven the animals over the cliffs and recorded in considerable detail the methods by which this kind of hunting was done.

Clark also mentions that he walked on the shore and "saw the remains of a number of buffalow, which had been drove down a Clift of rocks," but goes no further in attributing the cause of these deaths. This method of killing bison was one of the few available to people without guns and horses, and it was an ancient practice among the plains Indians. Lewis proposed that "in this manner the Indians of the Missouri distroy vast herds of buffaloe at a stroke." However, some experts familiar with the method and the area near the Judith River suggest that the "broken country back of this bluff is not really suitable for concentrating and stampeding buffalo" and therefore the large number of dead buffalo, which Clark estimated to be about one hundred, was more likely due to drowning during spring thaw and flood—due to changes in weather, rather than human actions.

This incident leads us to ask a question that has intrigued naturalists, ecologists, and anthropologists for decades: What was the relative impact of the Indians on bison compared to the effects of natural environmental change? In the nineteenth century, the famous British biologist Alfred Wallace wrote that an examination of the fossil record since the end of the ice age suggested that the "biggest and hugest and fiercest" animals, such as the saber-toothed tiger and the hairy mammoth, had died off. Some speculate that the changing climate at the end of the ice age was the cause. But these extinctions occurred around the time that the Indians were migrating to North America from Asia.

Paul Martin, an American anthropologist, has suggested instead that these extinctions may have been due to hunting by the newly immigrating Indians. The Indians would have been an introduced predator whose methods were unfamiliar to the native animals, which might have had little fear of human beings. Martin has suggested that a densely populated, moving wave of people coming down from the north could have used just this kind of method to kill vast numbers of the big animals, leading to their extinction. The matter, like the cause of the death of the bison that Lewis and Clark found near the Judith River, remains unresolved.

Lewis described in detail driving bison over cliffs: "one of the most active and fleet young men is scelected and disguised in a robe of buffaloe skin." This man then positions himself near the herd and the precipice. Lewis wrote that "the other indians now surround the herd on the back and flanks and at a signal agreed on all shew themselves at the same time moving forward towards the buffaloe." This causes the animals to stampede. Then the man disguised in the bison skin reveals himself to the animals as if he were one of them and runs in front of them to get them to stampede toward the precipice. Blinded by fear, the bison keep going and fall over the cliff. This man has to be careful not to be run over by the buffalo. "If they are not very fleet runers the buffaloe tread them under foot and crush them to death, and sometimes drive them over the precepice also," Lewis wrote.

Could the Indians have caused the extinction of such huge animals as the mammoth and the saber-toothed tiger with such methods as driving them over jumps, along with killing individuals here and there with bows and arrows? There is no doubt that the Indians had large effects on the native

Madison Buffalo Jump, Montana. Indian drove bison over the steep slope shown here near Three Forks, Montana. Men drove bison; women worked at the base, butchering carcasses. The jump may have been used by people for thousands of years. *Kelly Spears.*

animals, including bison. But it is my guess that the biggest impact was through alteration of habitat—in the case of the plains Indians, the frequent setting of fires, which would have improved the habitat for grass-eating grazers such as bison and made it poorer for woodland-feeding animals.

We know today that it is generally very hard to completely extinguish a species by hunting down and killing all the individuals. Such hunting can greatly reduce the numbers of a species, but it is nearly impossible to get the very last animal—especially if the tools available are stone arrow points and wooden bows and arrows, and the method of transportation is the human foot.

A much easier approach is to affect habitat. Most of the extinctions that modern technological civilization has brought about have occurred through habitat change, including physical alteration of the habitat and the introduction of exotic predators, competitors, and parasites. Our relative effect on the environment and its animals and plants is a major question for today, as it has been historically. The challenge is to sustain these wild living resources in much smaller habitats and despite much greater human population pressures.

With the coming of European technology and the introduction of the horse and the gun, and later with the invention of the train and telegraph, the potential to kill off the bison through hunting increased greatly and almost succeeded. As discussed in chapter four, there were many factors that contributed to the near-extinction of the bison, including a concerted effort to eliminate them because they were the primary food source of the Indians and because there existed American and European markets for bison hides.

The plains Indians obtained horses and guns much later than is often thought—only about fifty years before Lewis and Clark watched them hunting with guns—in the mid-eighteenth century. They adjusted to this new technology rapidly, using guns with great success and, some ecologists believe, killing bison at a rate beyond the capacity of that species to regenerate.

Elsewhere, hunting often came close to causing extinctions of other species native to North America, but did not. Typically, when a species is reduced to a very small number, it is both hard to find and no longer valuable as a commodity, so the chase is abandoned. This happened with the bowhead whale, hunted from 1840 to 1920 by Yankee whalers out of New Bedford, Massachusetts. Often the hunters, no longer able to find the few pockets that remain, believe that the species is extinct. This was the case with the sea otter and the elephant seal, both also hunted in the nineteenth century.

Whether hunting can extinguish a species depends on the percentage of the population people are able to kill. This raises the question—just as we asked for grizzlies—of how many bison there were and where they were found. We can make estimates based upon historic records, including the records of Lewis and Clark.

Many writers estimate the total number of bison to have been forty million to fifty million. In 1867, the estimates for the area between the Platte River and the Concho River in Texas ranged from fifteen million to fifty million. Although the range of our estimates is great, the impression remains the same: There were huge numbers of bison in the American West even as late as 1868, numbering in the tens of millions and probably totaling fifty million or more. Ominously, that same year the Kansas Pacific Railroad advertised a "Grand Railway Excursion and Buffalo Hunt."

By the 1880s, it was commonly believed that bison were on their way to extinction, but a number of individuals—acting independently—began to save bison, putting them on ranches or on Indian reservations. These individuals included Buffalo Bill Cody and some Sioux Indians. The number of buffalo probably dropped to a few thousand, perhaps four thousand, at its minimum. The figure has increased greatly since then.

SCENES OF VISIONARY ENCHANTMENT

THE UPPER MISSOURI

A CURIOUS CHANGE occurs on the Missouri River upstream from Fort Peck Dam and from there to Loma, near Fort Benton, Montana. The river you can see at Fort Benton or on a float trip through the White Cliffs area, the wild and scenic portion of the Missouri River, is very different from the river upstream from Great Falls or downstream from Fort Peck Dam.

Lewis and Clark saw this change in the Missouri River as they slowly moved upstream in the spring of 1805. On May 7, the expedition was just downstream from the location of modern Fort Peck Dam, where Lewis wrote that "the country we passed today on the North side of the river is one of the most beautifull plains we have yet seen." He saw that the land rose "gradually" away from the river "to the hight of 50 or 60 feet, then becoming level as a bowling green." That green landscape extended "back as far as the eye can reach." But the floodplain of the Missouri changed abruptly upstream.

Four days later, when the expedition was a little upstream of the dam site, Clark wrote that the "high land is rugged and approaches nearer than below, the hills and bluffs exhibit more mineral . . . Salts than below." Lewis's journal confirms this change with almost identical wording. Below the site

of Fort Peck, the Missouri flowed, as it does today, in a wide and gently sloping valley. Above it, the river flowed through a narrow and steep valley.
Today, this change is not visible in a single view, because the upstream portion at Fort Peck Dam is covered by water. But you can see the change when
you compare the Missouri River seen from the causeway below Fort Peck
Dam with the Missouri at Fort Benton, Loma, Judith Landing, or Virgelle,
Montana, and especially the famous White Cliffs section of the river.

You may wonder why the river is so different in these two sections. Perhaps it is simply because the upper Missouri section is closer to the headwaters and the Rocky Mountains, where the river still carries a heavy load
of sediment that can knife a steep edge into the countryside. But the Missouri upstream from Great Falls flows in a wider valley, not as steep as in
the White Cliffs section, so this explanation can't be right.

Ice-Age Missouri

A key to why the upper Missouri River is so different from the lower is
found in a visit to the Milk River, a tributary that flows into the Missouri
just below Fort Peck Dam. That river has a different look and a different
setting than the Missouri upstream from the dam.

The expedition arrived at the mouth of the Milk River on May 8, 1805.
Lewis saw the difference when he walked upstream along that river. He
wrote that "we nooned it just above the entrance of a large river," and continued that he "took the advantage of this leasure moment and examined
the river about 3 miles." He found the Milk River to be deep and gentle,
with a "large boddy of water." But most important, he saw that "the bottoms of this stream ar wide, level, fertile." He named it the Milk River
because the water had "a peculiar whiteness, being about the colour of a
cup of tea with the admixture of a tablespoonfull of milk."

Lewis and Clark observed each of the rivers they came across carefully.
They wrote down each river's characteristics and gave the rivers names,
true to a careful natural history examination. This is not the way most of us
view rivers in the countryside. Most of the time when we travel, we accept
the countryside as it is, as a static picture passing by us, without question-
ing how it came about. But rivers on a landscape are telling us a story—a
story about their history and why they are the kind of river we see. Now

Fort Peck Dam, near Glasgow, Montana. Said to be the largest embankment dam in the United States, the fifth largest man-made reservoir, the second largest volume embankment in the world. *U.S. Army Corps of Engineers.*

here was a curiosity: two rivers coming together on the same landscape at the same location, but one, the Missouri, had cut itself steeply between bluffs, while the other flowed in a wide and gentle floodplain. What caused the difference?

Geologists talk about "uniformitarianism," a big word that means that the processes that exist today existed in the past, and also that the processes that occur in one place occur in another—the physical, chemical, and biological processes that create a landscape have to follow the same rules of nature everywhere, in time and in space. If that is true, then the Milk and Missouri rivers ought to look the same and flow in similar valleys. How can these two rivers be so different?

It's a curious question, and the answer lies in the effects of the great continental ice sheets on the Missouri River tens of thousands of years ago. During the last ice age, the ice sheet pushed down from Canada all the way to the Missouri River in this part of Montana. The ice was an irresistible force, and the Missouri—that mighty river—was not quite the immovable object. The ice pushed the river out of its old bed along a section between Loma and Fort Peck, and then began melting back about fifteen thousand years ago.

The Valley of the Milk River. This is the old bed of the Missouri river, a bed of a mature, wide, and meandering river. The Missouri was displaced from this valley by the Pleistocene glaciers. The modern Milk River is small and narrow, too small to create this wide valley. *D. B. Botkin.*

When we see a great river, it appears to us as a permanent and unchanging part of the landscape. But a river has a history and goes through stages from youth to maturity. When a river first starts flowing on a landscape, it cuts straight down. The sediment it carries wears away at the land. A young river flows in a narrow valley with steep sides. But over a much longer time, the river keeps undercutting the bluffs along its narrow valley. The bluffs collapse. Lewis and Clark saw cliffs that had fallen into the Missouri just this way. The river then moves the debris from those fallen bluffs downstream. Slowly the valley widens. The slopes become gentle. When the valley is wide enough, the river can meander over it, and over the years it shifts its channel, creating oxbows, oxbow lakes, and backwaters. It becomes a mature river in a mature river valley.

The Missouri is an ancient river, and for most of its length it flows through the wide and gently sloping valley that characterizes such a river. But when the ice sheet pushed the Missouri out of its old bed, the river was forced to create a new one. During the height of the ice age, the Missouri was pushed south and forced to flow just to the south of the ice, where it began to cut a new valley into the countryside. Once that valley was formed, the river was captured by it.

When the ice sheet retreated, it left debris in the Missouri's old channel. The river had cut its way down into the new one and continues to flow through it today, from Loma to Fort Peck. From Great Falls to Loma, the Missouri flows through a wide valley. But because of glaciers long ago, at Loma the Missouri begins to flow through a narrow canyon and continues to do so past where Lewis and Clark noticed the change in the countryside.

The result is a landscape considered the most beautiful stretch of the Missouri. "The hills and river Clifts which we passed today exhibit a most romantic appearance," Lewis wrote on May 31, 1805, when they were in this white cliffs country, also known as the Missouri Breaks, upstream from the mouth of the Judith River. "The bluffs of the river rise to the hight of from 2 to 300 feet and in most places nearly perpendicular. . . . The water in the course of time in decending from those hills and plains on either side of the river has trickled down the soft sand clifts and woarn it into a thousand grotesque figures, which with the help of a little immagination and an oblique view at a distance, are made to represent eligant ranges of lofty freestone buildings, having their parapets well stocked with statuary: collumns of various sculpture both grooved and plain, are also seen supporting long galleries in front of those buildings."

It is remarkable that Lewis both appreciated the grandeur of this fanciful countryside and had insight into the geological processes that produced it. The first person of European descent to see this reach of the Missouri, he was also the first to understand how it could have formed through natural geological processes: "the thin stratas of hard freestone intermixed with the soft sandstone seems to have aided the water in forming this curious scenery."

Also, for one of the few times in the journey (another was when he arrived at the Great Falls a short time afterwards), Lewis let his imagination roam on the pages of his journal. He saw "long galleries in front of those buildings . . . ruins of eligant buildings; some collumns standing and almost entire with their pedestals and capitals; other retaining their pedestals but deprived by time or accident of their capitals; some lying prostrate . . . othe[r]s in the form of vast pyramids of connic structure . . . nitches and alcoves of various forms and sizes. . . . it seemed as if those seens of visionary inchantment would never have [an] end, for here it is too that nature presents to the view of the traveler vast ranges of walls of tolerable workmanship."

Perhaps most striking to us, Lewis wrote, "I should have thought that nature had attempted here to rival the human art of masonry had I not recollected that she had first began her work."

In 1834, Karl Bodmer painted this landscape, leaving us a record of how it appeared not long after Lewis and Clark passed this way and before European settlement. In this painting he shows several abrupt, sharp peaks with a particularly prominent one just left of the center of the painting and along the riverbank. In the distance, much lighter, almost white rocks form a line of cliffs. Beneath these is a tallus slope—a slope of land made up of material that has eroded from the bedrock forming the cliffs. Trees grow along the river, some perhaps cottonwoods; another up the slope just behind the highest pinnacle perhaps an evergreen. It seems a fanciful landscape, perhaps out of a myth—something that, today, one might expect to see in a film version of Tolkien's *Lord of the Rings*. Or perhaps out of a nineteenth-century romance.

This rock structure still exists within the wild and scenic portion of the upper Missouri. The major, central pinnacle towering over the river is easily

One of the Striking Formations Along the White Cliffs of the Upper Missouri River. Karl Bodmer, *Citadel Rock on the Upper Missouri*, watercolor on paper (JAM.1986.49.191). *Joslyn Art Museum, Omaha, Nebraska.*

identified, as are the off-white cliffs in the background and the darker, steep but gentler formations in front of the whitish cliffs.

The photograph suggests that perhaps Bodmer foreshortened the scene, making structures that were farther apart from left to right appear closer together and more abrupt. Otherwise, there is a great similarity in the scenes. Examining Bodmer's pinnacle, we can see that there are several delicate rock structures pointing upwards that do not appear today. Perhaps these existed in 1834 but later eroded, as such delicate structures tend to do. Bodmer shows birds flying low over the water, but no large trees on the bank opposite the pinnacle. The tree that appears in the year 2000 photograph would not have existed in 1834.

What we see from this comparison is that the White Cliffs section of the Missouri appears to have changed little in its geological formations. It is still fanciful, still recalling myth and legend or perhaps the remains of some ancient, long-passed civilization.

Part of the reason that the geological structures have changed little is the dry climate. In the modern city of Great Falls, Montana, the average precipitation is about fifteen inches a year. As mentioned earlier, this is less than the amount needed to sustain forests. In the Montana plains, much of the water for plants, wildlife, cattle, and cities comes from runoff from the Rocky Mountains. Thus the Montana plains depend on the snowpack in the Rockies. A snowy winter there means a good water supply for many parts of the Montana plains. Water, the great eroder, has much less opportunity to act here than in the wetter climates of Missouri, where Lewis and Clark began their journey, or along the Pacific coast, where they would soon be traveling.

The climate in Great Falls is cool and dry. Average maximum temperatures in the summer during the past thirty years have peaked at about 70 degrees Fahrenheit, and the average minimum has dropped to about 10 degrees F, with extreme highs reaching into the 90s and extreme lows in the minus-40s.

What, then, has changed on the upper Missouri River within the White Cliffs region? The changes are subtler than on the lower Missouri. Above Gavins Point Dam, the farthest downstream of the six major dams, the river was not channelized. It still runs free, more or less. In the White Cliffs areas, the river has embedded itself within steep slopes, and there is no room for it to meander—that will only come in ages far in the future, when the river has eroded a wide floodplain. There are three small dams upriver from the White Cliffs section, and these allow human control of the river's flow—to some extent, so that the seasonal variation in flow is less pronounced today than it was when Lewis and Clark were here. This could affect some seasonal habitats, where the steep slopes and incised river allow these habitats to form.

But the most important change in this part of the Missouri River is in the land along the river and back from it, land that is good for ranching. Much of this land is now grazed by cattle, which wade into the river, destroy riverside vegetation, and pollute the river with their droppings. Anyone who canoes this beautiful section of the river becomes well acquainted with manure. Landing for lunch or to make an overnight campsite sometimes requires stepping carefully around cattle pies.

As the ice sheet melted, it left debris of boulders, rocks, sand, silt, and clay everywhere, helter-skelter, mixed together, but also somewhat smoothed out. The plain that Lewis saw where the Milk River flows into the Missouri, at the site of present-day Fort Peck, was bulldozed by the glaciers to create the rolling but relatively flat countryside that Lewis found so beautiful. Huge boulders, dropped here and there by the glaciers, are markers and testimony to the powerful work done by the moving sheets of ice. They stand along the roadside to remind us of that awe-ful geologic history.

After the ice sheet retreated, the Milk River began to flow in part of the Missouri's old channel. A smaller river, it passes through a plain too big for it to have created. It is a young river in an old river's bed.

From Fort Peck Dam, you can travel to see the Milk River by going up to Nashua, Glasgow, or Malta and visiting the Hewitt Lake National Wildlife Refuge, which is on the big bend of the Milk River, or the Bowdon National Wildlife Refuge. When you do this, you are seeing the original valley of the Missouri. Just imagine a bigger river in this valley and you will be able to imagine the way this countryside would look today if there had never been the great climate change of the ice age, a time when ice hundreds of feet thick pushed aside the landscape in its path; scraped and eroded mountains and hills; and dumped the rocks and soil it had cut into valleys, disrupting the rivers and covering the forests. There is a great irony in this history—an environmental change that we would not want to happen now created great beauty that we want very much. It was an incredible large-scale change in the land—the arrival of a huge sheet of ice that persisited for thousands of years. We would do whatever we could to prevent the migration of the Missouri River into a new channel that is now considered its most beautiful stretch.

Which Is the Main Channel?

On June 3, 1805, the expedition was camped at what we know now as the mouth of the Marias River, but which appeared to Lewis and Clark as the junction of two large rivers. The problem was that they weren't sure which of the rivers was the real Missouri—the river that flowed in from the north, or the one that flowed in from the west. In a certain sense, this is arbitrary, because at the confluence of two major waterways, you can call either one the upstream continuation of the main river. But for Lewis and Clark, the question was, which river would take them the farthest into the mountains and give them the best route over the Rockies to the Columbia? The Indians had told them to search for a river that had some great falls on it. This would lead them to the trails used by the Indians to pass over the mountains. This is the river they would call the real Missouri.

Decision Point at the Marias River. The confluence of the Missouri and Marias Rivers. *D. B. Botkin.*

Choosing the wrong river and following it would have serious consequences for the expedition. Lewis wrote that "to mistake the stream at this period of the season, two months of the traveling season having now elapsed, and to ascend such stream to the rocky Mountain or perhaps much further before we could inform ourselves whether it did approach the Columbia or not, and then be obliged to return and take the other stream would not only loose us the whole of this season but would probably so dishearten the party that it might defeat the expedition altogether."

Although the future of the expedition, even perhaps their lives, depended on the right choice, Lewis took a detached, almost scientific approach, as if he were a modern scientist sitting in a comfortable laboratory office, rather than at a rough camp in bad weather. He pursued the problem with a seeming academic curiosity, writing, "An interesting question was now to be determined; which of these rivers was the Missouri."

In a sense, the expedition was lost and in need of directions. We have all been lost at some point or another. The trouble was, in Lewis's time, there wasn't anybody handy to give them directions. Today you can stop at a gas station, use your cell phone if you have one, or rely on a GPS device in your car.

The problem the expedition faced was one of their own uncertainty. The rivers were set in their directions and were not about to move at random over the next few days.

And so Lewis proposed an experiment: "to this end an investigation of both streams was the first thing to be done." He recognized the need to measure things about the river, making quantitative observations, to "learn their widths, debths, comparitive rappidity . . . and thence the comparitive bodies of water furnished by each," and by these means attempt to infer which was the main stream.

Like a modern scientific team, the camp divided into two groups, each examining the available evidence and each proposing what we would describe today as an hypothesis. Most of the men believed the north fork was the main river and therefore the one to follow. Lewis reviewed the evidence on their side: the north fork was deeper but not as swift. However, its waters ran "in the same boiling and roling manner which has uniformly characterized the Missouri throughout its whole course." The waters were brown, thick, and turbid—the Big Muddy, so it seemed. The bed of the river was also mainly mud, so that the "air & character of this river" seemed "precisely that of the missouri below." For these reasons, most of those on the expedition were convinced that the north fork was the Missouri. On the other side were Lewis and Clark, who, Lewis wrote, were "not quite so precipitate."

They decided to explore both forks, what scientists would call testing the two hypotheses. The next morning Clark led a group up the left fork, while Lewis took a group on the right. The rest of the expedition remained at the base camp where the two forks joined. Lewis traveled up the north fork from June 4 to June 6. He found that this fork continued northward toward what is now the border between Montana and Alberta, Canada, and became convinced that this path took them too far north to be the route to the Pacific.

After taking time to attempt a reading of the latitude and longitude (which failed because of cloudy weather), he began his return on June 7 to the junction of the two forks to rejoin the main body of the expedition. Lewis was correct: the north fork was a small tributary that they named the Marias River (actually Maria's River, in honor of Lewis's cousin, Maria Wood, but after a while, people dropped the apostrophe).

By spending a few extra days on the Marias River seeking measurements, he was trying to reduce the uncertainty about the position of the expedition. But a change in the weather, something he could not make accurate predictions about, prevented him from making the measurements.

Kinds of Uncertainty

In deciding which was the right river, Lewis and Clark were confronted simply with a lack of information. They were uncertain about what to do and wanted to avoid making a crucial error. The decision they faced at the junction of the two rivers involved what we can call *uncertainty of the first kind—a problem about the facts of a situation that already exists, or, given present conditions, must occur.* One of the channels was the main river—a fact that was not going to change during the time of the expedition. There was only one correct river to take. There was something direct and simple to do to resolve this uncertainty—explore the two rivers and determine by direct observation which was the correct one.

There is another quality of nature that leads to uncertainty about what we can do. Lewis experienced this quality on his way back to the confluence of the Marias and Missouri rivers: the problem of knowing that certain events can happen, but not knowing when. This is *an uncertainty of the second kind—uncertainty of the occurrence of some event that has some probability of happening, but whose occurrence involves inherent uncertainty.*

I call this the Las Vegas uncertainty: Will you place a bet on dice that haven't been rolled yet? Unlike the first kind of uncertainty, the second kind is not resolved directly and simply. You can't pick up your cell phone, call the weather bureau, and ask to know with complete certainty whether a thunderstorm will strike Loma exactly where the two rivers come together. The best a weatherman can do is give you the odds on whether or not it will happen. We cannot reduce the inherent uncertainty of this kind of future event by studying it. This is the problem with the flooding on the Missouri River. It is an uncertainty of the second kind that leads us to build levees and dams.

We can, however, learn the odds—or at least get an estimate—and decide if we want to accept them. We know, because people have rolled dice for a long time and also because mathematical analyses exist, the chances of rolling various numbers with a pair of dice, and we know that the number seven is the most likely sum to roll. But we can't find a sure route to seven, the way we can take Route 87 to Loma.

Perhaps our problem with this kind of error in our knowledge is a matter of relative time scales. Lewis and Clark spent more than a year on the Missouri, long enough for them to see and feel the river rising from storms or falling with changes in the seasons. Lewis and Clark often referred to changes in the water level, from the beginning of the trip until they reached the headwaters of the Missouri River. At Camp Dubois, before the trip began, the risings and fallings of the Mississippi were a persistent problem for Clark, who had to ensure that the boats were safe. On April 21, 1804, he noted that it rained the whole night and that the "river raised last night 12 Inches." On June 8, Clark referred to the high-water mark of the Missouri River. On April 22, 1805, when the expedition was in western North Dakota approaching what is now the Montana border, Clark observed that the river was "riseing a little."

In this day of satellite and aircraft observations, of automobile travel and vacations that are quick stops here and there, our ability to observe is curtailed. Most of us have just one shot to see the Missouri River, if we have any. If it is flooded, well, we may lose that chance. If it's a dry year, we will remember the river as it looked that year, and it will be fixed in our imagination as if it were always and forever that way.

On Lewis's return, as he was following the Marias River downstream, it began to rain, and the peculiar clay soil of the floodplain turned into a slippery mess, difficult to traverse. After a "most disagreable and wrestless night" camped in the rain, Lewis and his small band set off downriver on June 7 to join the rest of the expedition. The clay soil prevented the rain from soaking through and became so slippery that it was like "walking over frozan growund which is thawed to small debth." We know today that they were walking on a clay derived from glacial till and shale that is commonly called gumbo, a clay that turns into a plastic and sticky material when wet.

Lewis slipped on this soil while walking on a bluff above the river but managed to save himself from falling ninety feet into the water. Just after he had saved himself, he heard one of his men, Windsor, "cry out, good god Capt. what shall I do." He saw that Windsor had slipped on the clay and slid so that his right arm and leg hung over the bluff and that he was holding onto the edge with his left arm and leg. "I expected every instant to see him loose his strength and slip off," Lewis wrote, but "I disguised my feelings and spoke very calmly to him and assured him that he was in no kind of danger." Lewis then astutely told Windsor to take his knife out of his belt with the hand that was hanging over the precipice and dig a hole in the bank for his right foot, and by such effort work his way up, which Windsor did. In that way he was saved.

Searching for the right fork of a river is inherently different than trying to avoid slipping on wet clay and falling into the water. The *second kind of uncertainties are referred to today as problems of risk, because the event has not yet happened and its occurrence has to do with inherent chance, or with processes whose causes, for all practical purposes, we cannot distinguish from true chance events.* Translated into human events, risk becomes a matter of prediction, forecasting, luck, and fortune, the latter two of which were also constant companions of the expedition.

Our modern environmental problems confront us with both uncertainty due to lack of information and uncertainty due to chance, and it's important that we understand which kind we are facing. We do not seem to have trouble accepting the idea that we ourselves might not know which river to take. But we have great difficulty understanding and accepting the second kind of uncertainty—that there may be some inherent chance in nature.

If you have the time to take a canoe trip through the wild and scenic portion of the Missouri, you may have a chance to experience the river at a Lewis-and-Clark pace. When a friend of mine did this, he got caught in an intense thunderstorm and experienced the uncertainty of nature directly. Another friend canoed the region slightly upriver from Loma. He was caught in a strong easterly wind, a headwind, so he and his companion had to canoe hard even though the river was flowing with them.

The Missouri River's refusal to stay put and stay constant has been the source of many a good story and pithy saying, but this quality has also caused problems for homeowners, farmers, businesses, and conservationists. Most of our past methods of conserving and managing environmental factors have assumed the constancy of nature—except for human intervention. But the reality is the other way around. We try to fix a natural, varying environment, believing that our interventions cause variations in an otherwise static environment. Like the fickle Missouri, nature itself changes at many scales of time and space. We have longed for and tried to create an environment that is fixed, like the channelized Missouri downstream. Having lost our heritage about the river and the prairie, we seem to have ignored its important message.

10

"PLEASINGLY BEAUTIFUL" AND "SUBLIMELY GRAND"

PATHWAYS TO THE MOUNTAINS

THE EXPEDITION continued up the Missouri, leaving behind the strange and fascinating rock formations on the steep cliffs along the river. The countryside continued to be beautiful, in ways unlike most scenery known in Europe and the eastern United States. Lewis wrote on June 13, 1805, that "from the extremity of this roling country I overlooked a most beatifull and level plain of great extent or at least 50 or sixty miles." He observed that within this plain "were infinitely more buffaloe than I had ever before witnessed at a view." He was just downstream from the location of modern Fort Benton, Montana.

Today, just as Lewis described it, you will see a rolling but rather level plain, now filled with cattle, rather than bison, grazing the land, but with the same general appearance, at least from a distance. Even here the river has incised itself within this landscape north of Great Falls, Montana, cutting through the level plain, so that traveling away from the river on a main road such as Route 87, you are not aware that one of the greatest rivers of the world is nearby. It isn't visible.

Rising out of this plain, Lewis saw "two curious mountains" that were "square figures," probably the buttes just south of Black Horse Lake that

View of Great Falls, Montana, Today. *National Park Service.*

you can see if you drive south on Route 87. Lewis describes these as having perpendicular sides rising to a height of 250 feet and appearing to be formed of yellow clay.

The Great Falls

On June 13, 1805, Lewis came upon the first of the great falls of the Missouri River, now at the site of Ryan Dam, north of Great Falls, Montana. He was traveling with four men—Field, Drouillard (one of the main hunters of the expedition), Gibson, and Goodrich—and he sent three to kill some game for meat and then join him and Goodrich at the river for dinner.

"I had proceeded on this course about two miles with Goodrich at some distance behind me whin my ears were saluted with the agreeable sound of a fall of water and advancing a little further I saw the spray arrise above the plain like a collumn of smoke which would frequently dispear again in an instant caused I presume by the wind which blew pretty hard from the S. W.," Lewis wrote. The water "soon began to make a roaring too tremendious to be mistaken for any cause short of the great falls of the Missouri."

Great Falls of the Missouri before Construction of Dams. *U.S. Geological Survey.*

This was a welcome sound, because the Indians had told them earlier that the true Missouri River, the river that would lead them as far into the mountains as possible, had great waterfalls on it.

Walking sixteen miles, he and Goodrich reached the falls at noon. There they were, a small party in the midst of a huge region that was unmapped and unknown to European civilization. Reading Lewis's accounts, I admired the energy and ambition with which he rushed to see a place of beauty, when the expedition was about to be confronted with one of their most difficult tasks: portaging their equipment around these falls, which would take them about a month. But this was not on Lewis's mind at that moment. He heard the sound of a great fall of water and rushed to see what he hoped would be a beautiful view.

Lewis wrote that as he neared this point, "I hurryed down the hill which was about 200 feet high and difficult of access, to gaze on this sublimely grand specticle. I took my position on the top of some rocks about 20 feet high opposite the center of the falls."

When Lewis descended the steep slope, he saw a double falls, one just behind and above the other. The second, which he wrote was "an even sheet of water falling over a precipice of at least eighty feet," formed, with the first, "the grandest sight I ever beheld." The second falls was especially beautiful, because "the irregular and somewhat projecting rocks below receives the water in it's passage down and brakes it into a perfect white foam which assumes a thousand forms in a moment sometimes flying up in jets of sparkling foam to the hight of fifteen or twenty feet and are scarcely formed before large roling bodies of the same beaten and foaming water is thrown over and conceals them," he wrote. The rocks appear to have been perfectly placed to break up the water most beautifully.

In the river below, he saw a "butment of rock" that "defends a handsom little bottom of about three acres" and that was "agreeably shaded with some cottonwood trees; in the lower extremity of the bottom there is a very thick grove of the same kind of trees which are small." The land was inhabited: he saw among the trees several Indian lodges "formed of sticks."

For most of the journey, Lewis and Clark had maintained a rather distant and professional tone in their notes. Once in a while, one of them would write that he had seen a beautiful prairie or a wonderful and amazing number of animals, but these expressions about the beauty of nature were usually brief and reserved. At the time of their expedition, a great change was taking place in western civilization concerning the idea of natural beauty. The romantic poets of England—Wordsworth and Coleridge especially—were writing that the wildness of the Alps, with their fearsome heights, cliffs, ice, and wind, were objects of beauty.

Only a few decades before, mountains had been perceived, as they had been since Greek and Roman times, as horrible places, unsymmetrical and therefore ugly, as I discussed in chapter one. Until Lewis reaches the Great Falls, a reader of the journals would hardly know that Lewis was aware of such a debate over the aesthetics of nature. But something happened to him at the falls, and he opened up and wrote at considerable length about his own wonder at the beauty of the scenery, in the style of his time.

At this first set of falls, Lewis saw a rainbow in the spray as the sun reflected off the water. This, he wrote, "adds not a little to the beauty of this majestically grand senery." And for once he sought within himself an ability to express the beauty of the landscape, not just its capabilities: "after wrighting this imperfect discription I again viewed the falls and was so much disgusted with the imperfect idea which it conveyed of the scene that I determined to draw my pen across it and begin agin, but then reflected that I could not perhaps succeed better," he wrote. He wished for "the pencil of Salvator Rosa," a seventeenth-century Italian landscape painter of wild and desolate scenes, and for "the pen of Thompson," an eighteenth-century Scottish poet who was one of the forerunners of the Romantic movement. He wished that he had a camera obscura—a small room in which images of outside objects could be projected through a small hole, allowing an artist to trace the images exactly.

The next day, June 14, 1805, Lewis reached several more of the falls and was most impressed with one that he called Rainbow Falls, which is now much altered by Rainbow Dam. This is "one of the most beatifull objects in nature," he wrote. Lewis spent some time trying to decide which of the two—the falls he had seen the day before or this one—was the more beautiful. He wrote that "at length I determined between these two": Rainbow Falls was *"pleasingly beautifull"* while the one he had seen the day before was *"sublimely grand."* These are the turns of phrase that were used by the Romantic poets and their predecessors to describe aspects of beauty. "Beauty" referred to the classical Greek and Roman ideal of symmetry and perfection in geometry. "Sublime" had come into fashion among the romantic poets to refer to the awe-inspiring scenery of the Alps.

Lewis was using phrases that would have been familiar in the aristocratic drawing rooms of England and in Jefferson's Monticello but that were rarely used by other explorers of the American West in Lewis's time or for decades after—perhaps not until the nineteenth-century landscape painter Thomas Moran reached some of the great scenery of the American West following the Civil War. (Moran so popularized the scenery of the West that he probably helped inspire the creation of our national parks.)

The same day, Lewis "set one man about preparing a scaffold and collecting wood to dry the meat." He sent a message back to Clark to start searching for the best location to camp at the base of the falls for the portage around them. A few days later, he would have a dangerous encounter with a grizzly bear. Three weeks after that, he would be searching for Indians to sell him horses and guide him over the Rocky Mountains, before winter set in. The expedition was at a crucial juncture. But that was put aside when Lewis looked at the falls, especially the first: "I hope still to give to the world some faint idea of an object which at this moment fills me with such pleasure and astonishment, and which of it's kind I will venture to ascert is second to but one in the known world."

Clark arrived at the falls on June 17, 1805. In contrast to Lewis, Clark remained true to his propensity to report directly and make quantitative measurements—the first step in the scientific process. "I beheld those Cateracts with astonishment," he wrote, describing "the whole of the water of this great river Confined in a Channel of 280 yards and pitching over a rock of 97 feet 3/4," with a mist extending "for 150 yds. down & to near the top of the Clifts" so that the "river below is Confined a narrow Chanl. of 93 yards haveing a Small bottom of timber."

Today you can visit the site of the falls, but they have been dammed for hydroelectric power. The road to Ryan Dam is about six miles north of Great Falls, Montana, on Route 87. There you need to take a right off the road and follow the signs. After about nine miles (a total of about fifteen miles from Great Falls), the road descends steeply to the riverside and a parking lot. It is a pretty spot and worth a visit, but it is not the great falls that Lewis described.

Giant Springs and Giant Hailstorms

On June 18, Lewis and Clark were approaching and scouting out the great falls on the Missouri River to prepare for the portage of their materials around the falls, which would occupy them from June 21 to July 14. Clark set out early and, after passing the second of the great waterfalls, came on "the largest fountain or Spring I ever Saw." He made an estimate that this was "the largest in America Known." He was correct: the giant spring has been measured to discharge as much as 389 million gallons a day, with more recent measurements of 174 million to 213 million per day. This is enough water to cover one to two square miles a foot deep every day! He wrote that "this water boils up from under th rocks near the edge of the river and falls imediately into the river 8 feet and keeps its Colour for 1/2 a mile which is emencely Clear and of a bluish Cast."

Eleven days later, on June 29, Lewis set out to see the same spring with the French Canadian hunter Drouillard. On his way, Lewis described the countryside as "a level beautiful plain for about Six miles." Lewis too concluded that the fountain, as he called it, was "the largest I ever beheld." More likely than Clark to dwell on aesthetics, Lewis wrote that "the hadsome cascade which it affords over some steep and irregular rocks in it's passage to the river adds not a little to it's beauty." But like Clark, he also made measurements, writing that the spring was about twenty-five yards from the river, "situated in a pretty little level plain, and has a sudden decent of about 6 feet in one part of it's course." He noted, as did Clark, that the water was "extreemly tranparent and cold; nor is it impregnated with lime or any other extranious matter which I can discover, but is very pure and pleasent." There was so much water moving so quickly out of the ground that Lewis observed "the water of the fountain boil up with such force near it's center that it's surface in that part seems even higher than the surrounding earth which is a firm handsom terf of fine green grass."

Today the giant spring is in a city park within an urban setting. In the twentieth century, Great Falls developed around the production of electricity from downstream Ryan Dam, whose reservoir flooded and now covers the great falls for which the city is named. At one time, Anaconda Copper had a large refining plant here that converted ore to metal, using the electric power from Ryan Dam. Great Falls is a combination of pleasant residential, tired-out industrial, and pretty riverfront. All the riverfront is public land, and there is a marina with walkways and picnic tables here and there.

Curious geological processes created Giant Springs. A formation of limestone called Madison limestone lies under most of eastern Montana. It was formed about 250 million years ago from the deposits of shells and other biological matter in the bed of an ancient sea. Since the formation of the Rocky Mountains, which began about ninety million years ago, rain and snow have soaked into the limestone where it is exposed on the slopes of the Little Belt Mountains. From there the water drains downward and then flows through openings in the limestone to the Great Falls area. Next, under pressure because the water starts at a high elevation, the water flows upward and out at Giant Springs. A fracture in this limestone allows the water to be pushed up. The spring is a giant artesian well. Flowing through limestone, the water dissolves calcium and magnesium, which it brings to the surface.

On their way to the Giant Springs, Lewis and Drouillard were "overtaken by a violent gust of wind and rain from the S. W. attended with thunder and Litning." They took shelter "in a little gully wher there were some broad stones" that Lewis thought he could use to protect his head from hail. They remained for about an hour "without she[l]ter and took a copious drenching," Lewis wrote.

At the same time, Clark was ascending the riverside, along the series of falls, so that he could retake some notes about the river that he had lost on his previous ascent. With him was Charbonneau, the French Canadian interpreter; his Indian wife, Sacagawea; her baby boy; and York, the only black person on the expedition. They too saw the black cloud coming from the west. Lewis wrote that Clark "looked about for a shelter but could find none without being in great danger of being blown into the river should the wind prove as violent as it sometimes is on those occasions in these plains." Clark wrote that he found a deep ravine with "shelveing rocks" where they took shelter. He put his guns and the compass under one of these rocks. "Soon after a torrent of rain and hail fell more violent than ever I Saw before," Clark wrote.

The intensity was so great that it "fell like one voley of water falling from the heavens" and produced a flow of water into the ravine where he and Sacagawea had taken shelter "with emence force tareing every thing before it takeing with it large rocks & mud." It was clear that they had to get out of the ravine, which was flooding rapidly. He took his gun in his left hand and used his right to help Sacagawea, who was carrying her baby. Charbonneau, meanwhile, was trying to pull his wife up. "Before I got out of the bottom of the revein," Clark wrote, "the water was up to my waste & wet my watch."

He estimated that by the time he reached the top of the ravine, at least fifteen feet of water had risen. Sacagawea's baby had lost his clothes, and she was wet and cold and "just recovering from a Severe indispostion." Clark was "fearfull of a relaps." Other members of the expedition also suffered from that storm as they moved materials on the portage. "Some nearly killed one knocked down three times, and others without hats or any thing on their heads bloodey & Complained verry much," Clark wrote. He gave everybody a little grog.

In the midst of the escape, he had left the expedition's large and best compass, which was a "Serious loss," he wrote. Fortunately, the next morning two of the men went to the falls and found the compass covered with mud and sand, but everything else, including a tomahawk, a shot pouch, some powder and balls, some moccasins, and the baby's clothes and bedding, were gone. These men found that the place where Clark had sought shelter the day before was "filled with huge rocks," Lewis wrote.

Such rare events not only threatened the lives of Clark, Sacagawea, and her baby, but also have caused major changes in natural areas, creating new channels in a river or clearings in a forest. These rare and seldom seen events can play a major role in the dynamics of life on Earth, resetting the ecological clock to start natural processes of restoration and recovery to which many species are adapted. It is well that we be aware of them though we rarely experience them.

Gates of the Mountains

On July 19, 1805 the expedition came to an area of impressive scenery north of modern-day Helena, Montana. That evening, Lewis wrote that "we entered much the most remarkable clifts that we have yet seen." They seemed to "rise from the waters edge on either side perpendicularly to the hight of 1200 feet." It was impressive and a little forbidding, he wrote, adding that "every object here wears a dark and gloomy aspect. the tow[er]ing and projecting rocks in many places seem ready to tumble on us." Lewis then discussed the geology of the location, as Jefferson had instructed him to do: "the river appears to have forced it's way through this immence body of solid rock for the distance of 5 3/4 miles and where it makes it's exit below has thrown on either side vast collumns of rocks mountains high." It was so steep that for more than a mile, there were only "a few yards in extent on which a man could rest the soal of his foot." Then Lewis noted, with a bit of understatement, that "it was late in the evening before I entered this place and was obliged to continue my rout untill sometime after dark before I found a place sufficiently large to encamp my small party."

The rocks were of many shades and hues, Lewis wrote, from black to "yelloish brown and light creem colourd yellow." Clark described the hills as being made up of "a dark grey Stone & a redish brown intermixed," reported that "no one Clift is Solid rock, all the rocks of everry description is in Small pices appears to have been broken by Some Convulsion." The snow-capped mountains were in view, so Lewis wrote that "from the singular appearance of this place I called it the *gates of the rocky mountains.*"

Today the land nearby the river is little developed, and even though the Gates of the Mountains now lies between the reservoirs of Holter and Hauser dams on the Missouri, this area looks much as Lewis and Clark saw it. Large areas are protected for conservation. The Montana Land Reliance has obtained conservation easements on almost five hundred thousand acres, protecting almost nine hundred miles of streams and riverbanks.

Lewis wrote on July 19 that "whever we get a view of the lofty summits of the mountains the snow presents itself, alto' we are almost suffocated in this confined vally with heat." Above the river the next day, Lewis saw Douglas fir and Ponderosa pine, a "scattering of timber on the river and in the valley." There were bighorn sheep, beaver, and otter. These are still found in this area today, and if you are lucky you will see one of them. The area also remains a prime habitat for many birds, including pelicans, gulls, bald eagles, mergansers, meadowlarks, osprey, loons, Canada geese, peregrine falcons, and turkey vultures: 118 species have been spotted.

Today tours on large, open-air riverboats run several times within the Gates of the Mountains Recreation Area (from Memorial Day until October). These pass between the cliffs where Lewis and Clark traveled. A tour lasts close to two hours and includes a stop at Meriwether picnic area, where it is speculated the expedition camped for a night. From there, hikers may choose to explore the Gates of the Mountains wilderness area, catching a later boat ride back to shore.

The power of the river and the even-more-powerful forces that created the surrounding mountains capture your attention when you visit this location. If you take a boat tour, you might experience the power of the river as it cuts through the mountainous canyons and feel the awe that men of the expedition must have felt as they saw the nearby canyon opening up to the valley in the distance and the snow-capped mountains beyond.

On the 19th, Lewis wrote that "The river appears to have woarn a passage just the width of it's channel or 150 yrds." Confronted with such an amazing landscape, one can't help wondering what brings us the mountains. Since the early nineteenth century, soon after the Lewis and Clark journey, geologists recognized the processes of mountain building and mountain erosion. But no one had an explanation for how this mountain building came about nor where the energy came from. The answer is one of the great discoveries of twentieth-century geology.

Continents Colliding

In 1914, Alfred Wegener, a German scientist, proposed a radical theory that continents moved—or drifted. Wegener had observed the similarity of fossils of animals and plants found on different continents and noted the parallelism of coastlines of Africa and South America. But at the time the theory was dismissed. No one could conceive of a source of energy for that process, and the idea contradicted dominant theories about the constancy of nature. As the understanding of radioactivity increased, it became clear that the decay of radioactive elements deep in the Earth provided a source of energy and that with intense heat and pressure the material forming the crust of the Earth could act as a semiliquid. Today this theory of plate tectonics is well accepted.

The term *tectonics* comes from the Greek word for carpenter or builder. And if the river is the painter, the continents are the carpenters. Mountains are created in the collision of gigantic continental plates in motion. The deep earth acts as a semiliquid, and the cooler, lighter continents float on the surface, shifting about over time. The "solid" earth on which we stand moves. Heated from below, the continents are to the rest of the Earth as skim is to a chocolate pudding. The depth of the continents is no thicker relative to the rest of the planet than the skim is to that pudding. When the huge continental plates collide, mountains form as mere wrinkles on the surface.

The plates move slowly, but not so slowly that the movement cannot be measured. The average rate is about 3.5 inches a year—108 inches in the two hundred years since Lewis and Clark passed by the Gates of the Mountains. So today the Missouri River is nine feet further west in terms of fixed longitude than it was when Lewis and Clark were at this site.

Mountains have formed whenever continental plates collided, and some mountains are old, including the Appalachians of Virginia, the home of Lewis and Clark. The Rocky Mountains Lewis and Clark confronted at Gates of the Mountains are comparatively young, too young to have been worn smooth by rivers like the Gallatin, Jefferson, and Madison, the rivers that form the Missouri. That geological youth meant that in 1805, the Rockies were high, steep, and rough, a challenge unexpected by the expedition.

Geologist Brian Skinner has written that "it is not just the continents that move, it is the entire lithosphere. The continents, the ocean basins, and everything else on the surface of the Earth are moving along like passengers on large rafts; the rafts are huge plates of lithosphere that float on the underlying convecting material."

The colliding continental plates that formed the Rockies began their mountain building here near the end of the age of the dinosaurs and at the beginning of the age of mammals, more than sixty million years ago. They are built from even more ancient deposits. The lighter, brighter-colored rocks are ancient limestones that geologists call part of the Madison formation. These formed in a seabed, from deposits of ancient seashells and other materials, in the Mississippian era, more than three hundred million years ago. At this time, shallow seas were common on many continents, and especially in North America. The darker gray rocks that Clark described are even older. They are Greyson shale, formed more than six hundred million years ago.

Before the Rocky Mountains began to rise, when dinosaurs roamed this country, most of Montana was coastline—near sea level or underwater, part of a shallow sea that covered two-thirds of the state. After a continental plate that had been in the Pacific Ocean collided with the plate that formed North America, several things happened: The colliding westward plate formed the land that is now Washington and Oregon, where Lewis and Clark were soon to go. The Rockies began to rise; the land to the east that had been seashore also rose above sea level; and the sea was forced to retreat eastward.

Before the Rocky Mountains emerged, there was no Missouri River; and without the Rocky Mountains, there would still be none. A river is a necessary consequence of a mountain range. Water must flow downhill, and as it does it begins to carry sediment and erode a path. Tributaries begin to come together and form a young river. The young river cuts steeply through the rocks. But just how a river will form depends on bedrock, climate, and the stresses and strains, the cracks and bends, to which the rocks have been subjected over their longer history on Earth.

And so at Gates of the Mountains you have a dramatic view, as did Lewis and Clark, of primary forces that bring us the landscape from which begins the river that drains one-sixth of the continental United States. If you are able to take one of the excursion boats through Gates of the Mountains, you can imagine that you are on the continental raft being carried on a journey into the collision of continental plates. This is part of the new view of our planet, one of constant motion at all scales, of all materials. In this way we are all on a great journey, symbolized by the travel of Lewis and Clark up the Missouri River through the Gates of the Mountains.

Three Forks of the Missouri

As the expedition reached the headwaters of the Missouri, traveling on the river became more and more difficult. The river was ever more shallow, and the men had to drag the canoes over the rapids. The river current, descending from the steep mountains, was so swift that it was impossible for the men to paddle upstream even where the water was deep enough for the canoes to float. It took great energy to advance the canoes with ropes and poles.

Lewis and Clark divided the expedition into an advance group that went ahead on foot to explore the river and decide the best route, and a main group that pulled and poled the boats upstream. The exertion became exhausting and dangerous. Charbonneau sprained an ankle hiking in the rough country. Sergeant Gass fell in one of the boats and injured his back and so could not help pull or push the boats. Lewis assigned him to the advanced party on land.

During the winter at Fort Mandan, the Indians had told Lewis and Clark that when they followed the Missouri River, they would arrive at a place where three smaller rivers came together and flowed as one downstream. Clark took a small group of men and headed upstream. On July 25, they arrived at this location, which he called "Three Forks." Lewis arrived with the main party two days later.

As they approached Three Forks, the men spent much time wading so that they could push and pull their canoes upstream in the shallow waters; it was clear that soon canoes would be useless. Horses were needed to transport the expedition and the materials through the mountains, but Lewis and Clark had not been able to bring horses with them on the narrow upper Missouri and had left them behind on the plains. Now they were searching for Indians of the mountains with whom they could trade for horses.

Finding Indians and Horses

On July 27, Lewis wrote "we begin to feel considerable anxiety with rispect to the Snake Indians. If we do not find them or some other nation who have horses I fear the successfull issue of our voyage will be very doubtfull or at all events much more difficult in it's accomplishment."

Lewis made every effort to search out and approach Indians in a way that would be accepted as friendly. But the mountain Indians were at war with neighboring tribes downriver, and they fled at the first sight of strangers. Earlier, on July 23, he had raised small flags on the canoes, hoping that Shoshone Indians would see them and understand that the party was not a group of Indians and therefore not enemies.

In reaching the headwaters of the Missouri, they had also reached the homeland of their Indian guide, Sacagawea. Sacagawea was a Lemhi Shoshone who was born about 1788; taken prisoner about 1800; and purchased by Charbonneau, a French Canadian fur trader, about 1804. Lewis and Clark had first met Charbonneau and Sacagawea at the Mandan village where the expedition had spent the previous winter. Both Charbonneau and Sacagawea accompanied them on the rest of the trip, serving as interpreters. On July 22, Sacagawea told Lewis and Clark that she recognized the country and that they were near her relatives, information which "cheered the sperits of the party."

Not far downstream from the three forks, the river passed through a narrow channel "hemned in by high Clifts," Lewis wrote on July 27. He climbed to the top of one of these cliffs, a "beautifull spot" where he "commanded a most perfect view of the neighbouring country." Below he could make out the three branches that converged, two meeting upstream and the third, the southeastern fork, joining the others a little farther downstream. Each passed for many miles through large green meadows—riverside wetlands and floodplains. Between the southeastern branch and the middle branch, he saw "a distant range of lofty mountains" with "snow-clad tops—" the mountains that would be one of their greatest tests, and that they hoped would provide a short route to the Columbia, were near.

Rejoining the main party, Lewis found that the cliffs soon opened up. He passed the southeastern fork and followed the southwestern one for only one and three-quarter miles, where he set up camp. He wrote that "beleiving this to be an essential point in the geography of this western part of the Continent. I determined to remain at all events untill I obtained the necessary data for fixing it's latitude Longitude &c." They settled in, unloaded the canoes, and secured their goods on shore; and then several men went out to hunt.

Game was abundant and healthy. Lewis wrote on July 29, that "some of the hunters turned out and returned in a few hours with four fat bucks, the venison is now very fine," and he reported that the deer were all white tail, not the mule deer they had seen in the plains. He saw sandhill cranes in meadows, and "the hunters" caught a young one that Lewis found "very ferce" and "after amusing myself with it" let it go. There were trout (probably cutthroat), kingfishers, and ducks in the water, and grasshoppers, crickets, and ants that made "little perimids of small gravel in a conic shape, about 10 or 12 inches high," most likely the western harvester ant.

Lewis walked through the streamside meadows and examined the middle and southwestern forks, whose junction was upstream from where the southeastern stream joined the main river. Once again, a question that had caused the expedition considerable time came to the fore: Which was the Missouri? "I walked down to the middle fork and examined and compared it with the S.W. fork," Lewis wrote on the 27th, "but could not satisfy myself which was the largest stream." He decided that neither could be called the Missouri in preference to the other, because "they appeared if they had been cast in the same mould" and there was "no difference in character or size." Each was about ninety yards wide.

Soon after Clark rejoined the main body, having explored the southwestern branch some twenty-five miles above, during which he suffered from sunstroke and lack of water and was sick at the camp for several days. Reflecting on the similarity of these three branches, Lewis and Clark decided to call none of these the Missouri and to instead consider them separate rivers and give each its own name. They decided that the confluence of these three streams would thereby be marked as the headwaters of the Missouri. They named the southwest fork the Jefferson, the middle fork the Madison, and the southeast fork *"Gallitin's River"* in honor of Albert Gallatin, the secretary of the treasury.

In a sense, the decision not to call any of these tributaries the Missouri was arbitrary, as events of the next weeks would demonstrate. After several days' stay at Three Forks, they decided that the southwestern fork, the Jefferson, was most likely the river that would take them farthest west and into the mountains, and they chose to follow it to its headwaters. Once again the expedition divided into groups.

On July 30, Lewis walked ahead of the rest of the men, following the Jefferson River and seeing "a vast number of beaver in many large dams which they had maid in various bayoes of the river which are distributed to the distance of three or four miles on this side." He walked through "an extensive bottom of timbered and meadow lands intermixed," and had to wade through beaver ponds "to my waist in mud and water." He camped alone on an island, shot and cooked a duck for dinner, and kept warm by a fire.

Although Lewis wrote that wildlife was scarce, on August 1 he saw a herd of elk, two of which he and Drouillard killed; a few deer; and goats. That day, Lewis wrote that Clark "passed through the mountains" that "formed tremendious clifts of ragged and nearly perpendicular rocks" that he called a "black grannite" but that must have been limestone, since that is the kind of rock that exists there. It is three hundred million to more than four hundred million years old.

On that same day, Clark entered what Lewis called a "beautifull valley" more than six miles wide where the river ran "crooked and crouded with islands, it's bottoms wide fertile and covered with fine grass from 9 inches to 2 feet high and possesses but a scant proportion of timber, which consists almost entirely of a few narrow leafed cottonwood trees"—[narrowleaf cottonwood] found, as usual, along the river. The next day, Lewis wrote that the "land is tolerably fertile, consisting of a black or dark yellow loam, and covered with grass from 9 Inches to 2 feet high." He was somewhere between the modern towns of Cardwell and Whitehall, Montana.

On August 3, Lewis described the vegetation of the wide valley as lacking timber "except a scant proportion of cottonwood near the river" and possessing an understory of "narrow leafed or small willow, the small honeysuckle, rosebushes, currant, serviceberry, and goosberry bushes; also a small species of berch," which he noted had a "leaf which is oval finely, indented, small and of a deep green colour." Then in the evening, he "passed through a high plain for about 8 miles covered with prickley pears and bearded grass," a description that indicates the dryness of the region. In this valley, he saw "many deer, Antelopes ducks, gees, some beaver and great appearance of their work. also a small bird and the Curlooe as usual." That day "Reubin Fields killed a large Panther" that measured "seven and 1/2 feet from the nose to the extremity of the tail."

The Continental Divide and Contact with the Shoshone

Clark found the tracks of an Indian on August 3, but these suggested that the Indian had seen the party and run away. Lewis and Drouillard found an "old indian road" on August 5, a day on which Lewis estimated he walked twenty-five miles. By that day, Clark noted, the men were "much fatigued from their excessive labours in hauling the Canoes over the rapids."

One of the canoes overturned on August 6 and Lewis wrote, "all the bagage wet, the medecine box among other articles." In addition, "two other canoes had filled with water and wet their cargoes completely," wetting their cornmeal and many presents they had collected for the Indians. One of the men, Whitehouse, was thrown from a canoe, which then turned and "pressed him to the bottom as she passed over him," Lewis wrote, "and had the water been 2 inches shallower must inevitably have crushed him to death." Clearly, they could not proceed much further without horses.

On August 8, Sacagawea told Lewis that she recognized a peak that her people used as their summering grounds and that they called Beaver's Head. (This peak, located in Madison County along Montana Highway 41, about twelve miles southwest of the town of Twin Bridges, is known today by the same name.)

They persevered, crossing small streams and rough country until they reached what Lewis concluded was the very beginning of the Jefferson. There, on August 12, he stopped and drank the water: "judge then of the pleasure I felt in allying my thirst with this pure and ice cold water which issues from the base of a low mountain or hill," he wrote, for he had reached "the most distant fountain of the waters of the mighty Missouri in surch of which we have spent so many toilsome days and wristless nights. thus far I had accomplished one of those great objects on which my mind has been unalterably fixed for many years."

Others of the crew were equally joyful. Lewis noted that "two miles below McNeal had exultingly stood with a foot on each side of this little rivulet and thanked his god that he had lived to bestride the mighty & heretofore deemed endless Missouri." Lewis used that name for the first of the headwaters, which he had been calling the Jefferson. Having decided that this stream would take them the farthest into the Rockies and the closest to the Continental Divide, Lewis could have named the southwest branch the Missouri River. But he did not: the name "Jefferson River" remained. No matter, it is arbitrary. The feat, the struggle, the long and dangerous trip had accomplished its first major objective.

Having drunk from the Missouri's first water, Lewis walked up to the top of this eastern slope, crossing the Continental Divide—the location where all rivers to the east flow into the Missouri to the Mississippi and to the Gulf of Mexico, or north, east of the Rockies to Hudson Bay; and all the rivers to the west flow to the Pacific Ocean. He walked a short way down the western slope. "I now decended the mountain about 3/4 of a mile," he wrote, "to a handsome bold running Creek of cold Clear water. here I first tasted the water of the great Columbia river." He was drinking out of Horseshoe Bend Creek, a tributary of the Lemhi River, which in turns flows into the Salmon and Snake and then into the Columbia. He had crossed the Continental Divide, but he had yet to make contact with the Indians.

Lewis had seen an Indian on horseback on August 11, while traveling with his hunter, Drouillard, and Shields. Lewis believed he was a Shoshone because of his dress and his "eligant" horse, which he rode without a saddle. With his usual keen diplomacy, Lewis approached the Indian in a gentle and friendly manner. He made a standard signal of friendship by holding a blanket at two corners, "throwing [it] up in the air higher than the head," and bringing it to the earth as if to spread it. This was repeated three times. However, the two men with him did not have the same finesse and instead continued approaching the Indian, even though Lewis signaled them to stop. This caused the Indian to move away, although Lewis repeatedly called out "*Tabba-bone*," which he thought meant white man. Failing to make contact with this Indian, Lewis wrote that he felt "mortification and disappointment" and was "chargrined" by the behavior of his men, with whom he spoke about their failure. Continuing his careful planning, Lewis next attached some moccasins, some strings of beads, some paint, and a looking glass to a pole, which he inserted in the ground in hopes of attracting Indians.

On August 13, Lewis followed an Indian road. Several miles along it, he saw one man and two women. Once again he attempted to approach them with an unfurled flag, but the women fled and the man soon followed. About a mile farther, he came upon two girls and a woman, whom Lewis and his men surprised and approached so closely that the women apparently believed there was no escape and sat on the ground "holding down their heads as if reconciled to die." Lewis took the older woman by the hand to show her his skin. He then gave her some beads, a looking glass, and other small gifts. He painted the woman's face and the faces of the two girls with vermillion, which Lewis believed was an Indian emblem of peace. Then, using sign language, he asked them to bring him to their camp, and they complied.

In about two miles, they were met by sixty mounted and armed men, who embraced Lewis and his men according to their customs. Lewis spoke with the chief, whose name, he recorded, was Ca-me-ah-wait. He and his companions remained with this group for several days, but the Indians vacillated between friendliness, fear, and hostility. Lewis came to believe that the Indians thought Lewis was in league with their enemies and was leading them into an ambush. Finally, on August 16, Lewis gave his gun to the chief and told him that "if his enimies were in those bushes . . . he could defend himself with that gun," adding that he was "not affraid to die" and that he "might make what uce of the gun he thought proper or in other words that he might shoot me."

Lewis was waiting for the rest of the expedition to catch up with them and was concerned that the group of Indians might leave and "secrete themselves in the mountains where it would be impossible to find them." But the expedition joined them on August 17, and Sacagawea discovered that Chief Cameahwait was her brother. By amazing coincidence, the very first Indians they contacted turned out to be Sacagawea's own family.

On August 18, Lewis traded a uniform coat, a pair of leggings, three knives, some handkerchiefs, and other materials that he estimated were worth a total of twenty dollars for three "very good horses." So the two

Sacagaewa Meeting the Shoshone. Charles M. Russell, *Lewis and Clark Expedition*, 1918. Oil on canvas, 30.5 x 48.5 in. *The Thomas Gilcrease Institute of American History and Art, Tulsa, Oklahoma.*

things the expedition needed most desperately—horses and friendly guides—had been obtained. Good fortune, Lewis's astuteness and diplomacy, his and Clark's strength and courage, the hard work and cooperation of the men, and the assistance of Sacagawea all contributed to this important development. So it was that the greatest wilderness expedition ever taken in North America by people of European descent ultimately depended on courage, intelligence, technology, and the cooperation of the peoples of that land. As it would be on the return, the way west over the mountains was accomplished successfully because Lewis and Clark were traveling in the country known, lived in, and traveled through by Native Americans. The integration of people *and* nature was the foundation of their success.

Here, at a location now under Clark Canyon Reservoir, but then below the forks of the Beaverhead River, the expedition made camp with Sacagawea's relatives, and here they remained until August 24.

Ecological Transition

As they had traveled from the great falls upstream to the Continental Divide, the expedition passed through an ecological transition, from the Great Plains to the eastern slope of the Rocky Mountains. New animals and plants appeared on the landscape. On August 1, Clark shot a bighorn sheep—which they ate—and Lewis saw "a flock of the black or dark brown phesants," blue grouse, one of which they shot, examined, and described. It was a new species, "fully a third larger than the common phesant of the Atlantic states," Lewis wrote, and then set down the first scientific description of this bird. The same day Lewis saw "a blue bird about the size of the common robbin," whose call and behavior he described. It was the pinyon jay, and his description of this bird was also a scientific first. On August 3, Fields killed a mountain lion. The animals of the plains were in their past, behind them; the animals of the mountains were coming into view.

Lewis was wary of this change, because it meant a transition from the abundant big game of the plains—especially the buffalo—to the wildlife-poor forests and mountains. The wealth of wildlife in this country was in and near the streams—beaver and otter in great abundance, along with fish and water birds. On August 3, Clark noted that they saw "great numbers of

Beaver Otter &c. Some fish trout & and bottle nose." This change in big game abundance characterizes transitions from grasslands to forests throughout the world. Big-game animals are most abundant in grasslands, as on the Serengeti plains in Africa, and their numbers decline rapidly with forest cover, as they do in Africa when one travels west from the Serengeti to tropical rain forests. Forests, whatever else their beauty, ecological value, and economic worth, provide meager amounts of meat for people to eat.

Although they were entering the mountains, where forests usually dominate, they found few trees. Lewis wrote on August 1, 1805, that "the mountains are extreemly bare of timber," thus they were forced to hike "through steep valleys exposed to the heat of the sun without shade and scarcely a breath of air." The eastern slope of the mountains was in the rain shadow of the Rockies. A rain shadow occurs where moisture-full breezes from the ocean flow inland and are pushed upwards by the mountains. Rising, they

Lemhi Pass as Seen from the Clearwater National Forest in August 1965 on the Montana-Idaho state line. Here the Lewis and Clark expedition left the Missouri River drainage, crossing along this range of mountains into the Columbia River drainage on its way to the Pacific Ocean. *U.S. Department of Agriculture, Forest Service.*

cool, condensing their water, which is then released as rain and falls on the western slopes and mountain summits. The air, thus dried, descends down the eastern slopes and, sinking, is warmed and expands. It is now able to absorb moisture from the land. Dry itself, it makes the land below it even drier. The rivers and streams were fed by the snows on the summits, but the surrounding lower-elevation countryside was dry. As a result, the expedition forced their canoes upstream against strong water currents but hiked through dry country.

Not only did the mountains create a dry, tree-poor climate, but the Indians may aided in deforestation. Lewis wrote on August 4 that "the Indians appear on some parts of the river to have distroyed a great proportion of the little timber which there is by seting fire to the bottoms."

Their diet began to shift from meat of the plains to mountain fruits—berries and currants: "we feasted suptuously on our wild fruit particularly the yellow courant and the deep purple servicebury which I found to be excellent," Lewis wrote on August 2. Everything that was happening to them was influenced to a great degree by the natural history of the location, by the geological formations that influence the climate, by the vegetation that was in turn influenced by that climate, by the change in wildlife that was a result of the change in vegetation and the decrease in rainfall. The steepness of the streams, their rapid and dangerous currents, and the steep and rough country were the products of the ancient and great mountain-building events that began about ninety million years ago to form the Rocky Mountains and, as a result, to produce the Missouri River. Ancient geological and modern ecological processes combined to challenge the expedition with tough traverses, little water except in the streams, and less and less game.

Lewis understood the changes in the transition from plains to mountains. On foot and by stream, Lewis and Clark had developed their own understanding of the natural history of the Missouri River. That learning was now at an end, to be replaced by the harsh lessons of the mountains and the Columbia River to the west.

11

A PASSAGE STEEP AND STONEY, STREWN WITH FALLEN TIMBER

The Bitterroot Mountains

HAVING ASCENDED the Missouri River and found its source at the Jefferson River in the foothills of the mountains, the expedition next had to cross the Bitterroot Mountains, then and now some of the wildest and most difficult country in the lower forty-eight states. On August 10, 1805, Lewis was near where modern Montana Highway 41 crosses the Beaverhead River when he looked at the high and rocky mountains that confronted the expedition; concluded that they had already climbed a considerable altitude; and wrote, *"I do not beleive that the world can furnish an example of a river runing to the extent which the Missouri and Jefferson's rivers do through such a mountainous country and at the same time so navigable as they are if the Columbia furnishes us such another example, a communication across the continent by water will be practicable and safe. But this I can scarcely hope from a knowledge of its having in it comparitively short course to the ocean the same number of feet to decend which the Missouri and Mississippi have from this point to the Gulph of Mexico."* Even at this stage, Lewis hoped to find a useful northwest passage just as he was about to discover, as he traveled through the Bitterroot Mountains, that this hope was in vain.

On August 14, he asked Sacagawea's brother, the chief of the tribe, about the geography of the country,

> but I soon found that his information fell far short of my expectation or wishes. he drew the river on which we now are to which he placed two branches just above us, and which he shewed me from the openings of the mountains were in view; he next made it discharge itself into a large river which flowed from the S. W. about ten miles below us, then continued this joint stream in the same direction of this valley or N. W. for one days march and then enclined it to the West for 2 more days march, here he placed a number of heeps of sand on each side which he informed me represented the vast moutains of rock eternally covered with snow through which the river passed. that the perpendicular and even juting rocks so closely hemned in the river that there was no possibilyte of passing along the shore; that the bed of the river was obstructed by sharp pointed rocks and the rapidity of the stream such that the whole surface of the river was beat into perfect foam as far as the eye could reach. that the mountains were also inaccessible to man or horse. he said that this being the state of the country in that direction that himself nor none of his nation had ever been further down the river than these mountains.

He told Lewis to talk with an old man, Lewis did so, and the man told him that it was a twenty-day journey to the next encampment of white people.

Furthermore, to make this trip would require seven days of climbing over steep and rocky mountains, with nothing to eat but roots, and with a good chance of encountering fierce tribes. This would be followed by ten days of travel "through a dry and parched sandy desert," also without food, and another three or four days to reach land that was "tolerable fertile and partially covered with timber."

The old man was telling Lewis about the Salmon River, one of the most rugged rivers in the United States. Lewis abandoned plans to seek the most direct western route and hired the old man, known as Toby, to guide them through the mountains on a trail known by the Shoshone.

On September 9, the expedition camped along Lolo Creek, in the Bitterroot valley, near modern Missoula, Montana, and below the imposing Bitterroot Mountains, over which they would soon have to cross. Lewis wrote that they called this location, in a pleasant plain near a stream, "Travellers rest." The site is now a state park. On their way to this camp, they had followed what they named the next year the Clark's Fork River, and which is still so named.

They were still in dry country, as indicated by Lewis's observations that "the country in the valley of this river is generally a prarie" and that "the growth [of trees] is almost altogether pine principally of the longleafed kind," what we call Ponderosa pine—a pine of dry, open country, commonly in savannah, the pines scattered among grasses. Today the scenery along the highway down into the Bitterroot valley resembles what Lewis and Clark saw: dry, open woodlands of grasses and pines. The Lee Metcalf National Wildlife Refuge in the Bitterroot valley south of Missoula, Mon-

Near Traveler's Rest, Montana, Where Lewis and Clark Stayed on Both their Out-Bound and Return Journeys. Looking west toward Lolo Pass along the Lolo Trail. *U.S. Department of Agriculture Forest Service.*

tana, is a good location to see the vegetation and wildlife of this part of the Lewis and Clark trail.

On the way to this campsite, Lewis described the country along the Bitterroot River as "generally a prarie and from five to 6 miles wide." He wrote, "the growth is almost altogether pine principally of the longleafed kind [ponderosa], with some spruce [Engelmann] and a kind of furr resembleing the scotch furr [probably Douglas fir]. near the wartercourses we find a small proportion of the narrow leafed cottonwood [probably black cottonwood] some redwood honeysuckle [western trumpet honeysuckle] and rosebushes form the scant proportion of underbrush to be seen." On September 7–9 Clark reported the countryside below to be "Snow top mountains to our left" with "bottoms as also the hills Stoney bad land" with some pines along the creeks and on the mountains. The soil was white gravelley. Clark noted that they killed four deer, four ducks, and three prairie fowl. The honeysuckle they saw was new to western science.

It would take them eleven days to reach the summit of these mountains, during which lack of game forced to eat their packhorses in the snow-capped higher elevations.

The travel over the Bitterroot Mountains was difficult, even though it was mostly along the Lolo Trail, a road probably in use by Indians for thousands of years. The going got difficult soon after they passed Lolo Hot Springs on September 13—famous then and now—on the way up from Travelers Rest. The next day, Clark wrote, "The Mountains which we passed to day much worst than yesterday the last excessively bad & Thickly Strowed with falling timber." They were moving into a forest of lodgepole pine, Engelmann spruce, subalpine fir, Douglas fir, grand fir, whitebark pine, and mountain hemlock, with each species' dominance depending on elevation, compass direction of a slope, and how recently the forests had been burned or damaged by a storm. It was "Steep & Stoney our men and horses much fatigued." They camped near modern Powell Ranger Station on the Lochsa River.

The next day was no better. They passed through more fallen timber on steep slopes caused, Clark speculated, by "fire & wind." Finding time to be a naturalist, he noted that these disturbances had "deprived the Greater part of the Southerley Sides of this mountain of its gren timber." Indeed, in modern forestry and ecology, it is well known that southern slopes in the northern hemisphere are sunnier, drier, and therefore more likely to burn than north facing slopes. But this would have been known to few even among those familiar with the outdoors in the early nineteenth century.

The going was so difficult that horses slipped and rolled down slopes. Some were hurt. One "which Carried my desk & Small trunk," Clark wrote, "Turned over & roled down a mountain for 40 yards & lodged against a tree, broke the Desk." The horse was "but little hurt," however.

Even worse was the view. Clark wrote that "from this point I observed a range of high mountains Covered with Snow from S E. to SW with Their top bald or void of timber." It was hardly the landscape envisioned by Arrowsmith, the mapmaker with whom Lewis had talked in St. Louis during the winter of 1803–1804, and who had said that the western mountains could be no more than "3520 Feet High above the Level of their Base." Instead, it was a mountain range of fire- and storm-damaged trees within mountains pushed upwards by dynamic plate tectonics. They were now on a route parallel to modern Highway 12, the main road from Missoula, Montana, to Lewiston, Idaho, and one of the best modern ways to see, from a comfortable distance, the terrible mountains over which Lewis and Clark passed.

The next day, September 16, Clark wrote, it "began to Snow about 3 hours before Day and Continud all day." They were forced again to travel through fallen timber. "I have been wet and as cold in every part as I ever was in my life, indeed I was at one time fearfull my feet would freeze in the thin mockersons which I wore," Clark wrote. This is one of the few times in the entire journey that he admits to suffering from the environment, and uses the word "fearfull." Still, he never says he himself is "afraid."

And so it continued—camping on snowy, wet ground; running low on food; finding little to hunt at high elevations and therefore shooting and eating colts they had procured from the Indians. On September 18, Clark admitted—highly unusually—that the spirits of the men were dampened. But he did get a view west to the distant Camas and Nez Perce prairies beyond the mountains, viewing them from what is now called Sherman Peak.

Finally, on September 19, after they had traveled six miles, "the ridge terminated," Lewis wrote, "and we to our inexpressable joy discovered a large tract of Prairie country lying to the S. W. and widening as it appeared to extend to the W." Still, the difficulties were not over: "the road was excessively dangerous along this creek being a narrow rockey path generally on the side of steep precipice, from which in many places if ether man or horse were precipitated they would inevitably be dashed in pieces," Lewis continued. Such is the formation of these mountains: with valleys so steep that boulders fallen from above fill the land along the streams so that there is no floodplain in the sense of a flattened, water-deposited land along the streambed.

But once again, in spite of the greatest difficulties and suffering, Lewis continued his observations of nature, writing a detailed description of the varied thrush *(Isoreus naevius)* and several other birds, including Steller's jay, which had not been described before. Lewis's observations are detailed to the point that he recorded the bird's call as "ch_-ah,ch_-_h. And he also noted, as he continued to travel through forests, that "the soil as you leave the hights of the mountains becomes gradually more fertile. the land through which we passed this evening is of an excellent quality tho very broken, it is a dark grey soil."

Lolo Pass Looking East from Packer Meadow, the Montana–Idaho state line, in the Clearwater National Forest. Bitterroot Mountains, over which Lewis and Clark passed, are in the background. May 1965. *U.S. Department of Agriculture Forest Service.*

Again, Help from Indians

On reaching the final summit on September 20 and looking down on what is now the Clearwater River valley in Idaho, Clark wrote he was in "butifull Countrey for three miles," where he saw many Indian lodges. One mile from the lodges, he saw three Indian boys who ran from him and hid. He handed his rifle to one of his men, got off his horse, and "found [two of the boys, and] gave them Small pieces of ribin and Sent them forward to the village." Then one of the men of the village "Came out to meet me with great Caution & Conducted us to a large Spacious Lodge."

The story, handed down from generation to generation of Nez Perce and recounted by Zoa Swayne in her book *Do Them No Harm*, is that it was a hot day and most of the able-bodied men were out hunting. Only a few scouts remained in the village as lookouts for enemies, along with older men, women, and children. Three boys of about six or seven took their small bows and arrows and headed to the mountains, where they came upon ground squirrels and began shooting at them.

They became so involved in this activity that they did not notice Clark and his men approaching until they heard the sound of horses. They hid behind tall grass, thinking that the strangers had not seen them. They saw men whose faces looked like buffalo hide because of their beards. The man in front handed a strange-looking stick to one behind him. Then he "opened a pack on his horse and, with a shiny knife, cut off three strips of red material about as long as a water snake. With these fluttering in his hands, he headed straight for the grass where the boys thought they were hidden." He handed a fluttering red strip to one of the boys and with sign language asked him to go to his village and tell people to come. The correspondence between the details of the Nez Perce folk stories and the notes by Clark tells us that the events were accurately remembered over the generations.

The next day, September 21, Clark wrote that the people "were glad to See us & gave us drid Sammon." Clark and his men camped at the village. He asked the chief about the path of the river, smoked a pipe with him, and gave him a handkerchief and a silver cord.

They were still in the rugged country, for on September 21, Lewis wrote that "we passed a broken country heavily timbered," where there were again many downed trees. On that day he also admitted, highly unusually, "I find

myself growing weak for the want of food and most of the men complain
of a similar deficiency and have fallen off very much."

But on that day, they camped on the Clearwater River, nearing the end
of the worst of their passage. They camped near modern Orofino, Montana, at a location that is now much altered by the Dworshak Dam on the
North Fork and by a fish hatchery. The next day, the expedition finally
crossed the Bitterroots, and Lewis wrote that "the pleasure I now felt in
having tryumphed over the rocky Mountains and decending once more to
a level and fertile country where there was every rational hope of finding a
comfortable subsistence of myself and party can be more readily conceived
than expressed." He wrote this as they were still riding through rough
forested, difficult country.

That same day, Clark and his men joined Lewis and the rest of the party
at a camp at a second village. "The planes appeared covered with Spectators
viewing the White men and the articles which we had," Clark wrote, but his
men were weak from lack of food and from the effects of eating roots to
which they were not accustomed. Some Indians stole a knife, a compass,
and other articles from one of the men.

According to the Nez Perce stories, there was an old woman named Wat-
ku-ese in the first village who was the only one who had seen white men
before, and those white men had cured her of an illness. She was pleased to
see the seven men of Clark's advance party. Then she heard that there were
more white men coming and that all of them would camp at the next village. According to the stories, Wat-ku-ese could not sleep. . . . These [white]
men faced great danger. With all the warriors gone on the warpath her people would feel threatened by the numbers. The seven who first appeared
were objects of curiosity; but great numbers of men were a threat. "Wat-ku-
ese knew that swift death could come to every stranger in the silence of the
night. A man, armed only with a kopluts [skull-cracker], could slip among
sleeping people and deal a death blow without a struggle."

Wat-ku-ese knew that these white men were friends, but her people
would not know this. She decided she had to get to their camp as soon as
she could. She rode her horse to their camp, becoming exhausted in the
heat of the day, exhausting her horse, and had to walk a part of the way.

As she feared, the Nez Perce were becoming afraid of the expedition.
They wondered why a Snake Indian, Sacagawea, had come with them, since
the Snake Indians were their enemy. They were frightened by the black

man. They began to talk about killing the members of the expedition when they were asleep, fearing that otherwise the white men would kill them. But Wat-ku-ese arrived, exhausted and faint, and fell asleep in a tepee. After a time, she came out and called to the men whom she had heard plotting to kill the white men. These are good men, she called to them. "Men like these

A Rock Outcrop Called the Devil's Chair, Along Lolo Trail, with beargrass in the foreground, shows the difficult country through which Lewis and Clark traveled. It is now in the Clearwater National Forest. July 1963. *U.S. Department of Agriculture, Forest Service.*

were good to me. Do not kill them. Do them no harm! Do them no harm!"
And with those words, she died.

There is no record in the journals that Lewis and Clark were aware they
were in danger, although they reported having some minor troubles with
the Indians. Can we accept this story as true? The similarity of the various
accounts of Lewis's encounter with the young boys suggests that the Indian
stories are reliable. If the rest are as accurate, then from these stories we
learn how much the expedition depended on the Indians' friendliness , as
well as on the paths and roads they created, the food they bartered for with
them when they were near starving in the mountains, and the horses they
obtained from them.

From the time that the expedition left the Missouri River to the time they
put boats into the Kooskooskee, a tributary of the Columbia, they had traveled 340 miles, "200 miles of which is a good road, 140 miles over a tremendous mountain, steep and broken." The mountains they had crossed are
formed of ancient granite and metaphoric rock, including gneiss and schist,

View of Typical Terrain Along the Lolo Trail. September 1955. *U.S. Department of Agriculture, Forest Service.*

possibly more than one billion years old, but forced upward about fifty million years ago as part of the tectonic building of the Rocky Mountains.

Finally, they set up a camp on the Clearwater River, across from the mouth of the river's north fork and about five miles west of Orofino, a site they called "Canoe Camp," now a park commemorating Lewis and Clark's stay. Here the expedition remained until October 7, building five canoes, each burned and cut from a tree. On September 27, Clark wrote that Lewis and "nearly all the men" were sick, "Complaining of ther bowels, a heaviness at the Stomach." By October 1, Clark noted that Lewis was "getting much better." They were sick from eating camus root.

On October 7, they put their canoes on the water and headed down the Clearwater, which would take them to the Snake River, which led to the Columbia. The passage through the mountains was over; the challenge of

Indian Post Office, Bitterroot Mountains, August 1965. Indian Post Office rock cairns on the Lolo Trail. These cairns were established by the Nez Perce Indians before the time of Lewis and Clark's travel over the Lolo Trail. They indicated the point to leave the Lolo Trail to travel south to good hunting and fishing on the Lochsa River. *U.S. Department of Agriculture, Forest Service.*

Clearwater National Forest, Looking Toward Bitterroot Mountain Range from Indian Post Office Lake, Along the Lolo Trail. July 1963. *U.S. Department of Agriculture, Forest Service.*

the Columbia River watershed and the weather of the Pacific coast's winter still lay ahead.

So it was, on their return the next summer, that Lewis and Clark knew there were Indian roads and Indian guides, and that there were deep snows in the summits of the mountains. They also knew it was the Indians and their established ways that would get them through.

The Trip Back—Over the Mountains Again

The return over the mountains was as difficult as the outward journey. On June 17, 1806, Lewis and Clark approached the summit of the Rocky Mountains once again in an attempt to return home after more than two years travel. They had fulfilled most of the goals of their journey as requested by President Jefferson. They had found a path across the continent. They had met tribes of Indians along the way and told them of the Louisiana Purchase, signed in 1803, granting all of this land to the new United States of America and giving the Indians their new "father" in Washington. They had noted the customs of these tribes, and rendered what they could of their languages into written script. They had observed the animals and plants of the countryside and written about the conditions of the soils, rivers, mountains, and minerals.

At the summit, they found snow twelve to fifteen feet deep and no clear way through. For the first time, they were forced to retreat. Turning back "melancholy and disappointed," the members of the small band retraced their hard-won steps downslope and westward. Lewis had directed each of his three sergeants to keep a journal during the expedition. The evening of November 3, Sergeant Gass, one of the hardiest and most important members of the group, wrote, "there was not the appearance of a green shrub, or anything for our horses to subsist on, and we know it cannot be better for four days march, even." He continued most curiously, "could we find the road," a task which appeared almost impossible, without a guide.

After their discussions on the lonely summit on June 17, 1806, expedition members began to retreat toward the previous day's camp, a place Lewis and Clark had named "Hungry Creek." Reaching that creek, they traveled two miles upstream until they found "some scanty grass" for their horses and camped for the night. The grass was "so scant" that their horses wandered far in the night looking for forage, and some were lost "a considerable distance among the thick timber on the hillsides."

On that June day, Lewis and Clark were searching for the road they had taken on their westward travels, an Indian road that had brought them through one of the most crucial parts of the entire trip, the traverse of the mountains between the Missouri and Columbia rivers. There, as elsewhere, their success depended on the experiences and aid of the Indians and the changes the Indians had wrought on the landscape. Lewis and Clark were traveling in *their* wilderness, but they were also traveling in the Indians' backyards, in the Indians' hunting and gathering grounds, along paths familiar to Native Americans, paths that took them from their homes for one season to their homes for another and from the lands of one tribe to another's. In that wilderness was a road familiar to the Indians, one on which the survival of the expedition ultimately depended. Lewis and Clark were traveling in a culturally different countryside. The Indians had lived with and within natural changes, scenic grandeur, lack of symmetry. It was not the wilderness of the Jeffersonian imagination, nor of our contemporary imagination. It was the reality that was as an important scientific result as anything else Lewis and Clark discovered.

The next morning, June 18, the expedition did not find all the horses, though they searched from dawn until 9 A.M. Lewis and Clark sent two of the men to find the Chopunnish Indians, who had promised to provide a guide through the mountains, offering a rifle as a reward. The rest of the members began the upward journey once again. They crossed Hungry Creek, where the rocks and waters caused them great difficulty. One horse fell and threw its rider onto the rocks. Another man cut a vein in his leg and   the journal record reads "we had great difficulty in stopping the blood." Food was scarce and difficult to hunt in the "thick underbrush and fallen timber."

They camped again after traveling only two miles uphill from where they had been on the fifteenth. They were low on food and salt, and they could find little to eat, for fish and deer were scarce, and they were bothered by mosquitoes. The next day, they retreated downhill once again. On June 21, 1806, Lewis wrote "we all felt some mortification in being thus compelled to retrace our steps through this tedious and difficult part of our route, obstructed with brush and innumerable logs of fallen timber which renders the traveling distressing and even dangerous to our horses." Thus they struggled for six days.

On June 23, they found two Indians who agreed to guide the expedition over the mountains. On the next day, the expedition began its second major attempt to find a way back across the Rockies. By June 26, Lewis and Clark had only reached the point from which they had turned back nine days before. Going onward, through snows still ten feet deep, they "crossed abruptly steep hills . . . near tremendous precipices, where, had our horses slipped, we should have been lost irrecoverably." They camped above the headwaters of streams, on a south slope where the snows had melted, finally finding grass that could be reached by their horses.

On June 27, the expedition reached a summit where there was a cone-shaped stone about six feet high with a pine pole on top, a cairn made by the Indians to mark the height of land. From this location, Lewis and Clark recorded that they had "a commanding view of the surrounding mountains, which so completely inclose us that, though we have once passed them, we almost despair of ever escaping from them without the assistance

Lolo Creek Today. *U.S. Department of Agriculture, Forest Service.*

of the Indians." In what they called "this trackless region," their guides never hesitated and were so accurate in the travel that "wherever the snow has disappeared . . . we find the summer road."

Two days later, they descended below the snows, finding a good camp for their horses to graze, deer for themselves to eat, and warm springs where they and their Indian guides bathed. They had crossed the main divide of the Bitterroot Mountains, passed from what is now Idaho to what is now Montana. Although the great divide of the Rockies lay ahead, they had left behind this great ordeal. Soon after, on July 1, they agreed to split into two groups and explore different routes eastward, and by July 3, Lewis's group had found a route "so well beaten that we could no longer mistake it" and left their Indian guides. Two days later, they found themselves in "an extensive, beautiful, and well-watered valley nearly 12 miles in length," and on July 7th, they reached the divide between the Columbia and the Missouri. The danger was passed. They had been led on the road through their wilderness by Indian guides and had survived.

Today you can cross the Bitterroot-Selway Mountains on a two-lane highway. Even traveling at sixty miles an hour in an automobile, you will wonder when you will finally come to the last peak, finally reach the summit of this impressive range of mountains. You may want to stop at Lolo Hot Springs, about twenty-five miles west of the town of Lolo, where the expedition camped and enjoyed the waters, or picnic or camp at Traveler's Rest, just a few miles south of town.

12

ROLL ON, COLUMBIA, ROLL ON

DOWN THE SNAKE AND COLUMBIA RIVERS

Thousand years. All this here water just a going to waste.

—WOODY GUTHRIE

DESCENDING THE Clearwater to the Snake, and the Snake to the Columbia, the expedition reached the now-famous Columbia River Gorge, one of the most beautiful river valleys on Earth. The gorge confronted Lewis and Clark not only with great beauty but also with new challenges. Clark wrote on October 18, 1805, that they "proceeded on down the great Columbia river," where the water was "bordered with black rugid rocks." After traveling "16 miles from the point the river passes into the range of high Countrey," he saw in the distance "a mountain bearing S. W. Conocal form Covered with Snow"— Mount Hood, considerably downriver. On a day clear enough to see that mountain from where Clark stood, the scenery would have been beautiful: a rugged landscape of black rocks contrasting with the pale, dried grasses of the wide valley; distant mountain ranges covered with darker tree-greens; and the snow-capped volcanic peak far in the distance.

The next day, Clark went ashore and "assended a high clift about 200 feet above the water from the top of which is a leavel plain extending up the river and off for a great extent." It afforded "a pros[pect?] of the river and countrey below for great extent both to the right and left." On October 24, they canoed "21/2 miles" from their camp of the previous night, "a tremendious black rock Presented itself high and Steep appearing to choke up the river," Clark wrote. Looking nearer, down at the river, he reported, "the water of this great river is compressed into a Chanel between two rocks not exceeding *forty five* yards wide continues and for a 1/4 of a mile when it again widens." He added, "The whole of the Current of this great river must at all Stages pass thro' this narrow chanel of 45 yards." This channel was very narrow compared to the river sections above and below.

On October 25, the expedition took the canoes through these rapids, an event that brought a "great number of Indians viewing us from the high rocks under which we had to pass." This was the cascade that the Indians had told them was "the worst place in passing through the gut." Lewis and Clark decided to portage as much as they could but decided also that the largest canoes could not be portaged and would have to be floated through. "I deturmined to pass through this place notwithstanding the horrid appearance of this agitated gut Swelling, boiling & whorling in every direction," Clark had written the day before.

Clark did not simply get the men into canoes and push out from shore. On the contrary, he carefully examined the route and "had men on the Shore with ropes to throw in in Case any acidence happened at the Whirl." Clark dealt with this wild reach of the Columbia as he had other dangers throughout the journey: not by assuming nature is orderly, but by observing its wildness carefully, accepting it, and making plans accordingly. He had adapted his mind and heart to the American West as it really was: harsh, demanding, uncaring of people. And in this case formed by Miocene basalts, part of the world's largest lava flows.

Clark was concerned about the violence of the water, but, as was usual with him, he was also thinking about the causes of it—the "compression" of the waters. "This rapid I observed as I passed opposit to it to be verry bad interseped with high rock and Small rockey Islands, here I observed banks of Muscle Shells banked up in the river in Several places," he had written on October 19. As bad as these were, the expedition was able to float canoes through them. A week later Clark was to find worse rapids, now known as The Dalles of the Columbia River, which would pose one of the great challenges of the Columbia. How would they get their equipment below the rapids and still reach the Pacific coast before winter?

On October 25, Lewis and Clark examined the rapids carefully. They walked "to See the place the Indians pointed out as the worst place in passing through the gut," Clark wrote. Since, he added, "portage was impractiable with our large Canoes, we Concluded to Make a portage of our most valuable articles and run the canoes thro accordingly on our return divided the party Some to take over the Canoes, and others to take our Stores across a portage of a mile to a place the Chanel below this bad whorl & Suck." To increase the likelihood of safe passage of the canoes, he had some of the men stand on the rocky sides of the channel with ropes "fixed on the Chanel" to help "any who Should unfortunately meet with difficuelty in passing through." This attempt to canoe down the cascades was an unusual sight, and Clark watched "great numbers of Indians viewing us from the high rocks under which we had to pass."

Once again, Lewis and Clark prevailed. Clark wrote that "the 3 firt Canoes passed thro very well, the 4th nearly filled with water, the last passed through by takeing in a little water, thus [we passed] Safely below what I conceved to be the worst part of this Chanel." A little later, "one of the canoes almost ran up on a rock, but the portage was successful, with no men or equipment lost.

Following Jefferson's instructions to report on the condition of the countryside, in spite of the great difficulty the expedition had just experienced, Clark recorded in his journal, "This Chanel is through a hard rough black rock, from 50–100 yards wide. Swelling and boiling in a most tremendious maner." Thus the expedition made its acquaintance with the Columbia River Gorge.

Below the rapids at The Dalles, the Columbia River widened and became "a butifull jentle Stream of about half a mile wide," Clark wrote on the 25th. They had passed the famous rapids. Immediately below them, Clark saw marine mammals, which he called sea otters but which were probably seals or sea lions, suggesting that they were now in the great estuary of the Columbia River.

The "face of the Countrey" along the river, Clark wrote, "above and about the falls, is Steep ruged and rockey open and contain but a Small preportion of erbage, no timber a fiew bushes excepted." They were on the dry side of the Cascades in what is now eastern Washington and Oregon, at the site of the modern town of The Dalles.

Indians Fishing Just Below Celilo Falls. It was at locations like this that the Indians told Clark they could take "Salmon as fast as they wish" and where Clark saw the water "Swelling and boiling in a most tremendious maner," and had to determine how to get his men, equipment, and canoes down the river. *U.S. Army Corps of Engineers.*

The Geological History of the Columbia River Gorge

Clark had prevailed through the first of the great challenges of the Columbia River—the passage of The Dalles. The falls were there as a result of ancient geological events only describable in superlatives.

The Columbia River, like the Missouri, is one of the greatest rivers of North America. But it is a very different river from the Missouri. It is shorter, flowing 1,200 miles compared to the Missouri's 2,600, but it carries much more water because it flows through some of the rainiest, wettest regions of the United States: its flow is more than six times the Missouri's, with an average drainage of 160 million acre-feet compared to the Missouri's 25 million. It drains a smaller area, including parts of British Columbia, Canada, and seven U.S. states: Oregon, Washington, Idaho, Montana, Nevada, Wyoming, and Utah.

From a geological perspective, both rivers are products of the great plate-tectonic movements that built the Rocky Mountains and the Cascades, through which the Columbia flows, as well as the still-dynamic earth along the Oregon and Washington coasts, where a string of active and semidormant volcanoes stand in almost a straight line, marking a geologically active zone of a long history—from Mount Lassen in California to Mount Ranier just south of Seattle.

Between forty and twenty million years ago, volcanic eruptions spread ash and lava over thousands of square miles, in perhaps thousands of flows, creating what is believed to be the largest lava flows on Earth. David Alt and Donald Hyndman, in *Roadside Geology of Oregon,* refer to these as a "swarm of large basalt dikes." The floods of basalt are beyond our imaginations. As Alt and Hyndman put it, this volcanic activity, through central and eastern Oregon, "involved eruption of enormous floods of basalt, some of which covered thousands of square miles with single lava flows having volumes measurable in hundreds of cubic miles," and "no such overwhelming eruptions of basalt have happened anywhere in the world during historic time so we have no eyewitness accounts to help us picture what they were like."

What were they like? Hot lava, like the lava one sees moving across the big island of Hawaii, might be our best present-day image, expanded to cover as far as the eye could see—only no eye, except in an airplane, could survive within those thousands of miles to watch. Through this great contortion, the landscape was formed, and Clark guided the expedition through only a tiny piece of these gigantic flows! Such was the balanced, harmonious, peaceful, Garden of Eden world imagined by Arrowsmith and Jefferson's mentors: molten rocks beyond our imagination creating a twisted landscape through which one river valley, and only one, could cut through the Cascade Mountains to the sea, creating one and only one passage to the north coast of the new United States.

The Columbia River and the landscapes through which it flows are the product of a geological history of catastrophes, of which the huge basalt flows are only the most recent. If any region of the Earth is characterized by dramatic changes, it is this one during the last two hundred million years. Prior to that, during the Paleozoic era, much of the region was under a sea. In what is now eastern Oregon, mountain ranges formed whose remnants are the Blue Mountains, a rural countryside of forests and small settlements, of picturesque landscapes punctuated by steep mountains—a region south of the path of Lewis and Clark and unknown to them.

Huge geological events began in the Triassic period, when volcanoes rose to form island arcs. In the Jurassic, the time of the dinosaurs, huge areas were filled by molten granite. Again in the Cretaceous, about seventy million years ago, much of Idaho was filled with an intrusion of granite.

Although the region appears to be a continuous sheet of continent, it is a connection between two of the Earth's great continental plates, and the collision of these has created a violent, contorted, twisted, blocky landscape for tens of millions of years. In the Eocene era, beginning about fifty million years ago, a mountain range formed that today is the coast range near Coos Bay—on the Oregon shore south of the Columbia River. The collision of the continental plates began this section's ring of fire, still active, with the rise of volcanoes in what is now the western Cascades. The great basalt floods began in the Miocene era fifteen million years ago, and these occurred until the last few million years. During the past three million years—about the time that our species evolved—eruptions of volcanoes were followed by ice ages. Huge ice-age lakes formed from debris and ice dams, which then broke, creating immense floods of unimaginable magnitude.

Together, these large-scale events, extending over a time and space difficult if not impossible for us to imagine, created the landscape through which Lewis and Clark floated, through storms and among rock islands and cascades, on their way from the Rocky Mountains to the Pacific Ocean. The dangers they confronted and over which they prevailed were not only a product of these immense past forces, but a mere shadow of them. They came, in geological terms, in a time of relative calm, *after* the basalt flows that would have been impassible to men, after the repeated ice-age lake floods that would have inundated them and their small crafts. And so it is within this violent, ever-changing landscape that we must view their journey and understand how they survived and prevailed, and how all other life forms native to this region evolved and adapted to this landscape—a landscape not only of violent, damaging forces, but also of rocks forming rich soils, and of climates giving great rainfall, and an ocean teeming with life.

The Transition to Forests

The expedition continued down the Columbia River and soon entered an ecological transition from grassland to forest. This forested landscape was far different from the countryside to the east where the expedition first reached the Columbia. There, in eastern Oregon and Washington, the river flowed through dry grassland. The transition from dry grassland to forest occurred rapidly, especially compared to the gradual change from prairie to forest that the expedition had experienced over the many weeks traveling up the Missouri River. Lewis and Clark canoed through this transition and wrote about it, and it is readily visible today.

At the Columbia River cascades, Clark wrote on October 25 that "the face of the Countre on both Side of the river" was "Steep, ruged and rockey open and contain but a Small preportion of erbage, no timber a fiew bushes excepted"; the hills to the west, however, had some "Scattering pine white Oake & co." On October 31, the expedition came to a huge, perpendicular rock in a meadowland along the north shore of the Columbia River, about fifty miles upriver from modern Portland, Oregon. They estimated that the rock was "about 800 feet high and 400 paces around." Now known as Beacon Rock and still a famous local landmark, it is a huge volcanic plug believed to have been ejected and thrown across the gorge by an ancient volcanic explosion.

"The low grounds are about three-quarters of a mile wide, rising gradually to the hills, with a rich soil covered with grass, fern, and other small undergrowth," the journals record at Beacon Rock, while the mountains, which approached the river "with steep rugged sides," were "covered with a very thick growth of pine, cedar, cottonwood, and oak." This was a pleasant change to Lewis and Clark. "The Countre has a handsom appearance," Clark wrote on November 3, and on the next day describing, "I walked out . . . [and] found the country fine, an open Prarie for 1 mile back of which the wood land comence riseing back." On the edge of the prairie, he saw white oak; on the hills, spruce, pine (his name for Douglas fir—western hemlock was and is also a common species), maple, and cottonwood growing near this river; he wrote, "After being so long accustomed to the dreary nakedness of the country above, the change is as grateful to the eye as it is useful in supplying us with fuel." The forests began to provide them—as they have to people before them and after—with a major resource on which civilization has always depended: timber for fuel and construction.

Beacon Rock also marked the end of the dangerous rapids through which the expedition had traveled in their boats. Downstream, the river widens. Four miles downstream, Lewis and Clark estimated that the river

Beacon Rock. A 1,000 foot tall basalt rock, which is either the core of an ancient volcano or an ejecta from a volcano—a huge rock exploded from an ancient volcano. It is now within the Beacon Rock State Park, 4,650 acres along the Columbia River. There is a trail to the top and a beautiful view of the gorge. Lewis and Clark wrote about this striking formation. Ecologically, it marks the beginning of the Cascades forests, and the transition from the grasslands of eastern Oregon and Washington, the dry county with little rain, into the wet coastal region of the Pacific Northwest. *Courtesy of Washington State Park.*

was about two miles wide, and there they found "a Smoth gentle stream." They could feel their approach to the ocean, and Clark noted on November 2 that "the tide has its effect as high as the Beacon rock."

For anyone traveling through this region, Beacon Rock still serves as a good locator of a major environmental transition between the dry region of eastern Oregon and the wet coastal region. This transition is the result of a rain-shadow effect from the Cascades Mountains, through which Lewis and Clark were passing. As I mentioned earlier, a rain shadow occurs when a mountain range forces the air to rise as it passes inland from the ocean.

The rain-shadow effect is as dramatic today along the Columbia River Gorge as it was in 1805. In the short distance between Pasco, Washington (the confluence of the Yakima, Snake, and Columbia rivers), and Beacon Rock, the country changes climatic zones—from a desert to a temperate, wet coastal region. At Beacon Rock the total rainfall is not very high, the rain falls frequently as light drizzles, and occasional heavy storms move through. It is rainy and cloudy for most of the year, except during summer and early fall. The climate difference between eastern and western Oregon and Washington causes the change in vegetation. The Cascade Mountains form a major rain shadow, as do the coastal mountains to the west. Jefferson and his scientist mentors of the early nineteenth century did not understand that the western mountains would have such a great effect on climate over a short distance. This caused another lack of symmetry between the coasts. The prevailing westerly winds and eastward-moving weather systems in the Northern Hemisphere created an asymmetry in climate that was contrary to the Greek ideal. Once again, nature did not play fair with the old ideas of western civilization. Lewis and Clark would soon experience the full force of the stormy fall and winter as they traveled farther down the Columbia. Severe storms lay in wait while Lewis and Clark enjoyed the clear days and beautiful scenery of the eastern Cascades.

At The Dalles, Oregon, the average rainfall is about fifteen inches a year—the same as in the dry countryside of the Bitterroot valley where Lewis and Clark stopped at Travelers Rest and observed Ponderosa pine among grasslands. Farther west, at modern Hood River, the average precipitation is thirty inches a year, twice that on the east side of the Cascades. At Portland, the average precipitation is forty inches a year—similar to the amount of rainfall in Lewis and Clark's home states, but in a very different climate. While the precipitation along the east coast of the United States is more or less evenly distributed through the year, in the Pacific Northwest the summer is dry—generally rainless and hot—while the winter is cool and very wet. While the weather along the Columbia River was to pose difficulties for Lewis and Clark, today the great winds that blow through the gorge have made places like Hood River a world center for windsurfing, a sport unlikely to have been imagined at the beginning of the nineteenth century—not only because the technology was not available, but also because the idea that nature's violence and lack of symmetry could be enjoyed was just being realized. The Romantic poets, as Lewis knew and reflected upon when he tried to describe the Great Falls of the Missouri, had this new vision.

Salmon, Plausibilities, and Possibilities

On August 13, 1805, soon after he had crossed the Continental Divide, Captain Meriwether Lewis ate fresh salmon. The Indian chief Cameahwait, who Lewis was soon to discover was Sacagawea's brother, had brought him to a

Multnomah Falls is one of the most beautiful sights in the Columbia River Gorge. It is 620 feet high, the second highest perennial waterfall in the United States. The water flows down rock laced with leafy liverworts, mosses, and ferns, and a rich forest of Douglas fir and other species of the Pacific Northwest's temperate rainforest line the hillsides. It is said that more than two million people stop at the falls each year. *Diana Karabut.*

"bower" where he offered the roasted fish. Tasting the salmon, the first that he had seen during the fifteen months he had traveled from St. Louis, Lewis was satisfied that he was, at last, on waters that drained into the Pacific.

Lewis and the salmon, two very different kinds of travelers, had met. Each had traveled a long distance through many hardships. The salmon had swum nine hundred miles from the Pacific Ocean. It had entered at the wide mouth of the massive river, over its sandbars, against its mighty current. It had swum a hundred miles past the Willamette River and the site of modern-day Portland, Oregon, rejecting that large river, then jumped mighty falls at The Dalles on the Columbia, formed of ancient, massive lava flows. It had come to the confluence of the Yakima, Columbia, and Snake rivers. For reasons hidden only within itself, through some instinctive detection still not understood, it chose the mighty Snake. Others of its kind chose the Yakima or the Columbia.

It swam hundreds of miles farther, past still slopes of the massive lava layers, reaching another river divide, and choosing the Clearwater, a smaller but fast-running, clear river. On and on it went, unrelenting, not eating, living off the fat that it had accumulated from years of feeding on the rich, upwelling currents of the mighty Pacific Ocean. And in August of 1805, it had grabbed at a bait or been netted by Shoeshone. It had failed in its mission: to find its natal stream and reproduce.

During the lifetime of that salmon, Lewis and Clark had planned their expedition and traveled upstream against the six- to eight-mile-an-hour current of the Missouri. Like the salmon, their journey was filled with dangers and required many decisions. Like the salmon, they had passed many a tributary and been forced to choose which was the right stream for their mission: to find a northwest passage, the best possible water route from the Mississippi to the Pacific Ocean.

Today, if you were to travel to that same stream and go fishing, you would likely find no salmon; at best, a few still return that long distance, that nine hundred miles, to the small streams where Lewis dined with Cameahwait.

The parallel made here between human beings and salmon is not just a literary device. Salmon and *homo sapiens* share a long history. Fred Allendorf, professor of genetics at the University of Montana–Missoula, has traced that history. He writes, "*The temporal (evolutionary) dimension of life began on the Earth over 3 billion years ago. The evolutionary lineages leading to salmon and humans diverged from a common ancestor of most vertebrates approximately 400 million years ago. Thus, salmon and humans have shared over 80% of their evolutionary history. In the broad evolutionary view, we literally are the brothers and sisters of salmon."*

Obtaining salmon took much time and effort for the Indians, Lewis observed on October 10, when the expedition was along the Snake River near the Idaho–Oregon border and the site of modern Lewiston, Idaho. On that day, he wrote that "the Chopunnish have very few amusements, for their life is painful and laborious. . . . During the summer and autumn they are busily occupied in fishing for salmon . . . as the fish either perishes or returns about the first of September." He continued, "they are compelled at this season in search of subsistence to resort to the Missouri, in the valleys of which, there is more game even within the mountains." These Native Americans not only fished on tributaries of the Columbia, but traveled widely, moving across the Continental Divide to hunt in the plains to the east. For them, both the buffalo and the salmon were sustenance. Their repeated travels had created a road, and they had techniques to catch both kinds of game.

The incredible abundance of the oceangoing, river-breeding fish was observed by Clark on October 17, the day after the expedition reached the Columbia River. Local Indians gathered on the banks to look at the strange, bearded men. The Native Americans manned eighteen canoes and accompanied Clark and his men up the river. Clark saw salmon everywhere, swimming in the water, lying dead along the shore, and drying on scaffolds built by the Indians. The numbers were "almost inconceivable," he wrote. "The water is so clear that they can readily be seen at the depth of 15 or 20 feet; but at this season they float in such quantities down the stream, and are drifted ashore, that the Indians have only to collect, split, and dry them on the scaffolds." The great abundance of salmon continued for days afterward. On October 22, the expedition reached the Great Falls of the Columbia. There, on a small island, Clark counted 20 stacks of dried and pounded salmon." He also wrote that the Salmon are "put in very new baskets of about 90 or 100 pounds weight." But as we shall see, even this biological resource, capable of traveling one thousand miles up rivers and streams to spawn, had its limits.

The fish that Lewis and Clark saw once they had reached the western divide and began to follow the Lemhi, Salmon, Snake, and Columbia rivers to the Pacific Ocean included seven species: salmon–chinook, coho, chum, sockeye, pink, steelhead and sea-run cutthroat. These are known today as anadromous fish, meaning that they spawn, hatch, and rear in freshwaters; travel to the ocean, where they grow and put on most of their weight; and then return to the freshwater rivers to lay eggs. These seven species are closely related, belonging to the same genus.

The importance of salmon and the effort devoted to accessing this food source was made clear in the journals. The Indians on the Yakima River caught and dried salmon on scaffolds built of wood. In eastern Washington, the climate is dry, as discussed previously, because of the rain-shadow effect. There was little wood in the vicinity, and the Indians on the Yakima had to transport wood needed to process the salmon a long distance. Elsewhere, Clark saw that the Indians dried salmon to preserve the fish by burying them in the ground in straw.

From these accounts and others in the journals of the expedition, we learn that Pacific salmon were an essential part of the diet of Indians who lived over a large area, from the remote, mountain tributaries of the Salmon and Snake rivers in the mountains of Idaho, some three hundred airline miles from the ocean; to the shores along the Salmon and Snake; to the shores of the Columbia all the way to its mouth, where the expedition would spend the winter of 1805–1806. Thousands of miles of streams were involved. There were many difficult passages for the fish on their way upstream. The most difficult passage on the Columbia occurs at The Dalles, as we already know from Clark's descriptions.

The history of exploitation by people of European descent can be divided into five periods: First was a period of discovery that, for salmon on the Columbia River system, can be said to begin with Lewis and Clark. This period lasted for a half century after the expedition. The second, intense exploitation began for the salmon right after the Civil War, only beginning in earnest with the development of the Oregon Trail when people from the East began to migrate to the Pacific Northwest and settle there. The third period, of awakening conservation and the first attempts at professional, rational, scientific-based management of wild, living resources, began approximately at the end of World War I and continued until the 1960s. This was the period in which the goal of management was single-purposed—to maximize the production of a single resource for harvest. A fourth period of environmentalism began in the 1960s, with great growth in public awareness of environmental issues and with the decline of many wild, living resources. Although landmark legislation was passed in this period, the management of fisheries continued to follow the single-factor maximization of harvest that dominated the second period. We may be witnessing a fifth, yet-unnamed period, a period to which I hope this book will be a contribution.

Right after the Civil War, the catch of salmon on the Columbia increased exponentially, as new kinds of fish-catching devices were invented and applied. There were mechanical traps, driven by water power—the movement of the Columbia's waters downstream—that took the fish into a device resembling a ferris wheel and dumped them into holding tanks. There were line fisherman and net fishermen. Some describe fishing on the Columbia during the salmon runs as consisting of a series of mechanical trapping devices spread in a continuous line across the river.

Records of the catch on the Columbia have been kept since 1866, when a mere fifteen thousand chinook were caught by settlers on the river. As word spread about the incredible numbers of salmon, and as new harvesting machines were invented, the catch more than doubled each year for the first few years. In the second year, 1867, the catch more than quadrupled, to 66,000; that nearly doubled to 102,000 the next year, and more than tripled the next year to 367,000. In four years, from 1866 to 1870, the commercial catch of chinook had increased twenty times. As with many natural resources, the incredible initial, exploited abundance was so great that it seemed a resource to grab without care for the future. It was a classic example of exploitation of America's wild living resources, and it took place concurrently with the exploitation and demise of the buffalo.

Even this wonderful resource had its limits: the salmon have a limited capacity for production. The pattern of the harvest was like that of the buffalo—rising rapidly after the Civil War, peaking in a surprisingly short time, and then declining. The one major difference in these examples was that killing of the buffalo was for warfare against the plains Indians as well as for commerce, while the harvest of salmon was simply for commerce and sport.

Perhaps more interesting is that the time it took to reach that peak catch was similar to that for the buffalo: about two decades. At first, only chinook were taken in great numbers as a commercial catch. But the catch of chinook peaked in the early 1880s, declining after 1884. As the most desirable species declined, the next most desirable was taken, and so on through the list of the species. And as each species became a major commercial item, records were kept of its harvest. In 1889, the year that we first have records for commercial harvest of sockeye, 348,000 of these were caught; the same year 236,000 steelhead were taken. Coho salmon and even chum salmon, so called because they were at first considered too small to be worthy of harvest and later harvested packed in chunks or "chums," were added in the early 1890s.

The total catch of chinook, coho, sockeye, chum, and steelhead salmon reached 3.3 million fish in 1896. The catch rose above 3 million again in 1911 and 1915, reaching its greatest size, 3.6 million, in 1918. Harvests continued to range between 2.5 and 3 million until 1934. But a harvest of 2.5 to 3 million salmon was too much for the populations of these fish on the Columbia, and the new dams also began to take their toll. In 1934, the catch dropped to 1.4 million. The last commercial catch that surpassed 1 million on the Columbia occurred in 1948, except for one anomalous year, 1986, when there was a huge one-year increase in the catch of coho. By 1990, the total catch of all species had declined to 257,000, smaller than all the catches after 1869—smaller than the catch four years after the commercial exploitation began.

The salmon we catch are the ones returning to spawn; they have survived many dangers and most of their cohorts—the individuals of their stock born during the same year—have died. Salmon that spawn are not around to care for their young: either the adults are dead, as is usually the case, or they have returned to the ocean. Like many fish, salmon overcome lack of parental care by producing huge numbers of eggs, most of which are eaten or otherwise destroyed before they hatch, and some of fish die early in life. A stock of salmon may survive if, on average, at least two individuals from the eggs of each female—one male and one female—survive all causes of death to return to the stream to spawn again.

It is usual to discuss the growth of a population in any single year as the difference between the number born and the number that died, but so many young salmon die that the key measure is the number of adults that return to spawn at the end of each year compared to the number of adults that returned the previous year. For there to be a surplus population to harvest, this number must be equal to the number that spawned last year plus an excess population; it is this excess that can be harvested, leaving the population a reasonable chance to sustain itself. If the harvest is less than the excess, then the population may increase. Note that I wrote "may," not "will." The excess adults may not find a place to spawn. If the harvest exceeds the excess, then the population will decline. Note that I wrote "will," not "may." When the numbers of animals returning are as huge as they were in the nineteenth century, and when there are many factors that cause variations from year to year, it is hard to know whether the population is in an overall decline or just a one-year variation. It is just hard to tell a variation from a trend.

If the estimates for the numbers of adult salmon swimming up the Columbia in the 1920s and 1930s were more or less correct, then there would have been ten to fifteen million of them, and the annual catch would have represented 20 percent of the adult population. For the salmon to sustain that catch, their annual production of "excess" adults—adults surviving other causes of death, such as predation by seals and sea lions—would have had to equal this number. These adults return when they are three, four, and five years old. Enough young would have had to make the swim down the river and survive several years in the ocean that the population returning would be 2.5 to 3 million more than that required to produce enough young fish to replenish the population to what it was before the increase in harvest. And do the whole thing over again, and again, and again. As abundant and productive as these fish are, this is a highly implausible expectation for them, as it is for any medium-sized or large vertebrate. From my studies and the studies of many others, of the long-term histories of animal populations, this seems so unlikely that few would expect, plan for, or advise that such harvests be sustained.

It is amazing that salmon were able to sustain themselves at all on the Columbia or on most of the other rivers of the Pacific Northwest against the technology mounted against them: early photographs show devices of many kinds forming almost continuous lines across the river during upstream migrations. But salmon are adapted to episodic environments—to change, to variation. They are robust against these, and they have persisted. Amazingly, in spite of the onslaught against them, they have continued to number in the millions in the rivers of Oregon and Washington. Estimates for more recent times are 4.5 million fish swimming upstream in 1977, and between one and three million after 1983.

But these robust fish are not adapted to permanent alterations of their habitats that prevent their return. Floods, storms, droughts, forest fires, and their technological mimics—runoff variations, periods of water removal for irrigation, episodic logging—they can withstand. But dams without fish ladders that prevent a stock from reaching its spawning grounds for four or five years—enough so that no individual born there can return—end that stock. Permanent diversions for irrigation that keep the waters forever low and prevent spawning for four or five years also will end that stock. Such long-term or permanent alterations of habitat have, along with intense commercial harvest, decimated some stocks. Coho salmon, important on the Snake River at the time that Lewis and Clark came down that waterway, were eliminated from that river by 1986. So far, there has been no recovery: no known coho on the Snake River; none get to the Lemhi; none to the great Continental Divide.

What the Expedition Can Tell Us about Sustaining Salmon

Congress funded the Lewis and Clark expedition for a total of $10,000, roughly equivalent to $200,000 in 2005 dollars. *Ten thousand times* that amount, and more, has been spent in the last thirty years to try to save salmon. The Bonneville Power Administration, which built and administers the big dams on the Columbia and Snake rivers, has alone spent more than one billion dollars on salmon research and restoration, without a single sign of improvement, without a single gain in the numbers of salmon taking their incredible journey.

How could it be that the great expedition of Lewis and Clark succeeded with such a small expenditure of funds to explore a vast territory unknown to western civilization and unwritten about; that they discovered hundreds of new species and found their way to the Pacific and back, losing only one man; while today we seem unable to improve the net number of salmon making their journey? It has not been for lack of money or suggestions; nor for the lack of studies. Reports longer than the Bible accumulate, each trying to explain the demise of many stocks of salmon. Scientists debate and politicians promise; bureaucrats carry out instructions.

The story of the first meeting of Lewis and a salmon in the Bitterroot Mountains perhaps can help resolve this strange conundrum. Lewis and Clark understood how to approach, travel through, study, and survive in wilderness; they knew how to select among risks, how to make decisions with extremely limited information, with technologies that seem laughable against our modern satellite remote-sensing and geographic-positioning devices. Following Lewis and Clark through the Bitterroots, down the Clearwater, the Snake, and the Columbia; to the Pacific Ocean and examining their journals, we can gain insights about how we might solve problems like the demise of salmon. That journey, which was a journey of ideas and concepts as well as a physical travail, provides insights and guidelines. It was a matter of brains as well as brawn.

It is easy to find things to blame for the decline of salmon. Many have done so. One of the favorite, major culprits is the great dams on the Columbia River system. It is important to consider the reasons these dams were built and the evidence for and against the effects of the dams, relative to other causes, on salmon. This must be done acknowledging the environmental backdrop—the ecological play—of the Pacific Northwest, which, as we have seen, is a scenery of frequent, repeated, violent change. Salmon evolved, adapted, and persisted in this land, prevailing within many environments.

Dams on the Columbia

The Bonneville Power Administration constructed and operates eighteen major dams on the combined Columbia-Snake river systems, including their tributaries, extending all the way to the Brownlee on the Snake River, 609 miles from the ocean. The first of the great power dams on the Columbia was Rock Island, completed in 1933 in eastern Washington north of Richland, 453 miles from the river's mouth. The next, and perhaps the most famous, was the Bonneville Dam, 146 miles from the mouth of the river, completed in 1938 and built with fish ladders to allow salmon to pass.

Grand Coulee Dam, the third and largest in terms of its reservoir, was completed in eastern Washington in 1941, 545 miles upstream from the mouth. This dam created a huge reservoir, destroying many miles of spawning areas, and contained no passages for fish. One of the dams, The Dalles Dam, located not far upstream from the Bonneville Dam and above the present town of The Dalles, Oregon, flooded the famous cascades through which the expedition passed.

These dams have greatly altered the Columbia and Snake rivers, but in ways that are less obvious to the casual passerby than the dams, channeliza tion, and construction of levees on the Missouri River. In part, the effects of the dams are less evident because of the powerful geological structures that surround the Columbia and Snake rivers—the Cascade and Rocky mountains, the hard basalt rock flows. These have not allowed the rivers to meander throughout wide, river-created valleys, but constrained them between hard bedrock.

What have these eighteen dams done to the Columbia and Snake rivers? They converted the variable flows through cascades into a series of large reservoir-lakes. Although the rivers still run fast, they run at speeds slower than at the time of Lewis and Clark, and with quite different current patterns.

It is easy, and has become a tendency, to view the building of the great dams on the Columbia and Snake rivers as simply mistakes of the past, brought about perhaps by special interests against the wishes of the people and the needs of the land and its ecosystems. However, the dams were built by people who believed that these would do good, provide hope for those

Lower Granite Dam, Snake River. *U.S. Army Corps of Engineers.*

Construction of One of the Great Dams on the Columbia. *U.S. Army Corps of Engineers.*

who had suffered from the Dust Bowl and the Great Depression, create jobs, and improve American society. The Army Corps of Engineers phrased well the reasons for the creation of the great dams on the Columbia River system, writing, "From the beginning, Bonneville Dam has symbolized hope. In a time when jobs were scarce and hope was virtually nonexistent, Bonneville Dam meant a well-earned day's pay to thousands of men and women. It meant a rich rebirth for the communities nearby, and for an entire region badly in need of an infusion of federal dollars. As the dam grew out of the riverbed, foot by foot, day after day, it became a towering monument to the triumph of the working man over economic depression."

Woody Guthrie and the BPA

Today, when environmental issues seem to be so integral to our social and political landscape, it is difficult for some of us to imagine a time when other issues might have held an equal moral sway. But this may have been true in the 1930s, when the great dams on the Columbia and Snake rivers were imagined, designed, and begun. And this would also help explain why the Missouri River was channelized and dammed without public outcry. The evidence for this moral position is illustrated no better than by one of America's great folk singers, Woody Guthrie.

Woody Guthrie traveled and knew the Columbia. "I saw the Columbia River and the big Grand Coulee Dam from just about every cliff, mountain, tree, and post from which it can be seen," he wrote as mentioned briefly before. "I made up twenty-six songs about the Columbia and about the dam and about the men, and these songs were recorded by the Department of the Interior, Bonneville Power Administration out in Portland." There was a great public acceptance of Guthrie's ideas and songs. "The records were played at all sorts and sizes of meetings where people bought bonds to bring the power lines over the fields and hills to their own little places," he wrote. "Electricity to milk the cows, kiss the maid, shoe the old mare, light up the saloon, the chili joint window, the schools, and churches along the way, to run the factories turning out manganese, chrome, bauxite, aluminum and steel."

In the 1930s, engineering development of the Columbia River was part of radical chic, considered by a broad segment of our society to be a public good. To a large extent, attitudes today toward the environment fill the same role. Preservation of the environment has broad support socially, and a set of radical groups believe that the defense of the environment is so important as to supersede the well-being of people. To Woody Guthrie, the taming of the river was part of a radical political alteration of our society. To some environmental groups, the reverse action is an equally radical political alteration of our society. Woody Guthrie's dream did benefit the economic standard of living of the people; but the Bonneville Power Administration, his employer during that time, became a large, entrenched

Lock on the Columbia River. *U.S. Army Corps of Engineers.*

bureaucracy suffering from problems typical of such organizations—a focus on its own continuation, a lack of flexibility, a slowness to react. What had been radical chic became conservative establishment, unable to respond rapidly to the changing perceptions and needs of the time. Perhaps a major lesson from this history of the Columbia is that, in our enthusiasm to save the environment of the Pacific Northwest, we should not create huge bureaucracies that are unresponsive to changing and complex needs of the future.

What else can we learn from the difference between the idealized societal improvement envisioned in Woody Guthrie's songs and the actual development? We knew too little. Natural systems such as the Columbia River are much more complex than anyone realized half a century ago and than most recognize today. The real lesson is that, given the extent of our knowledge, we should be humble in our approach to both social and environmental engineering.

And then there are the salmon—with their intricate, complicated life cycle, many habitats, and highly specific needs. A great robustness has enabled this group of fish not only to survive, but often to prevail in spite of ice ages and volcanic flows that destroyed habitats for long periods. Compare what is necessary to save buffalo with what is necessary to save salmon. We can grow buffalo on rangeland much as we do cattle. Or we can put them into remnants of native or restored prairie where they can fend for themselves. That's comparatively easy. The buffalo are at one extreme in their requirements in the modern world; salmon, with their intricate requirements, at the opposite extreme.

A stream must have just the right combination of conditions for salmon to breed successfully and return year after year. If we want salmon in a particular stream, we must provide those conditions. But to make matters even more difficult, the "what" that we must provide is not a fixed thing or group of things, but a set of processes that proceed over time. The salmon have persisted not because there was a single stream that was their home forever, but because they were able to shift among many rivers and streams as climate and volcanic activity and forest fires changed the landscape. If we do not overharvest salmon, and if we instead allow streams and rivers to be available and enable at least some of them to have the right conditions at a given time, the salmon will persist.

We cannot provide this complex set of conditions if we insist that nature is static and unchanging. Here is the point at which a simplistic ideology, however appealing, just doesn't work—whether it is the ideology of Woody Guthrie's songs or the contemporary insistence that the streams and rivers be fixed in time and people be removed from the equation. If political pressures force a situation in which it is either salmon or people, the salmon will almost certainly lose.

13

CHANGING OLD FORESTS
AT THE MOUTH OF THE COLUMBIA

AFTER REACHING the mouth of the Columbia River in early December 1805, Lewis and Clark organized the men and built a wooden fort. They chose a site that was south of the Columbia along the wide mouth of an estuary, away from the terrible strong winds that poured down the Columbia from the interior mountains, and protected by coastal hills and forests from direct exposure to the storms that came in relentlessly from the Pacific. They named the site "Fort Clatsop" after the local Indians.

The weather at Fort Clatsop had a severe effect on the members of the expedition. They began to build huts on December 11, in the rain. On that day, several of the men suffered from "excessive dampness" and four had violent colds, one had dysentery, one had "tumors on his legs," and two had injured their limbs. Protection from the weather was desperately needed. On December 14, working again in the rain, they completed the walls of the cabins and Clark recorded that the constant rains spoiled "our last Supply of Elk" and that "Scerce one man in Camp Can bost of being one day dry Since we landed at this point." The insides of the cabins were finished on December 17, just in time, as their tents had rotted through and now tore at the slightest touch, and snow and hail fell the next morning, lasting until

Replica of Fort Clatsop and Its Surroundings. *Courtesy of National Park Service.*

noon, followed by rain, which continued until the next morning. The necessity of wooden shelters and wood fires was clear. On the day after Christmas, the journals record that the men attempted to dry their wet articles in front of fires.

Food was scarce throughout the winter. The heavy forest cover produced a dense shade near the ground, providing little forage for elk or deer and little for the members of the expedition to eat. The Indians survived on a diet of fish dried for the winter, big game, and native vegetation.

The dense forests had made obtaining food difficult well before the expedition reached winter camp. On November 11, Clark sent Joe Field out to hunt, but he returned shortly and said that the hills were high and steep, the undergrowth and fallen timber so thick that he could not proceed far.

On February 4, Lewis observed that "the Elk are in much better order in the point near the praries than they are in the woody country arround us or up the Netul. in the praries they feed on grass and rushes, considerable quantities of which are yet green and succulet. in the woody country their food is huckle berry bushes, fern, and an evergreen shrub which resembles the lorel in some measure; the last constitutes the greater part of their food and grows abundantly through all the timbered country, particularly the hillsides and more broken parts of it."

Dense Pacific Northwest Coastal Douglas-Fir Forest. *Jerry Franklin.*

During their stay at Fort Clatsop Lewis and Clark wrote frequently about the lack of food and the difficulty in finding game. On January 6, 1806, Clark left the fort with a group of men on one of the rare clear days of that winter. He wrote that "the evening a butifull Clear moon Shiney night, and the 1st fair night which we have had for 2 months." Twelve men went in two canoes. They pursued and killed one elk, which they ate entirely. During the first part of their stay at Fort Clatsop, they were able to find some deer and elk, but the animals were generally lean and the flesh by no means as good as what they had obtained earlier.

Replicates of the huts and the surrounding wood fort walls can be seen today at Fort Clatsop National Monument, about ten miles south of Astoria, Oregon. Reconstructed Fort Clatsop is a rectangle of huts made of roughly hewn logs not well chinked: the wind seeps through, chilling the air. These seem cruder than the houses of the Indians as Clark described them.

Reading Clark's accounts of the Indian houses and other wood crafts, one becomes aware that these inhabitants of the forest had the skills to fell big trees, craft large boards, and make many utensils from wood. This leads one to wonder if perhaps they had other skills that would give them power over their environment that today we tend not to attribute to such "primitive" people, such as an ability to modify the forest to enhance the production of food and structurally useful timber.

Identifying New Species

On November 18, 1805, Clark and eleven of the men walked to Cape Disappointment, in what is now the state of Washington, a place named by John Meares, an English trader who failed to cross the sandbars in the rough and complex currents at the mouth of the Columbia River. Interestingly, Meares made this attempt in 1788, the year that Jefferson wrote his book *Notes on the State of Virginia*. It is also interesting that Lewis and Clark had heard of Cape Disappointment, for it shows that exploration of the Pacific coast was known to them.

Clark found "the remains of a whale on the Sand." Unlike many of the other animals and plants Lewis and Clark saw, this dead whale was not subjected to careful observation or description.

Remembering that Lewis had been instructed by Jefferson to note "the dates at which particular plants put forth or lose their flower or leaf; times of appearance of particular birds, reptiles, or insects," one wonders why the whale was not included otherwise.

During their second winter, locked on the wet, almost sunless Pacific coast, they were surrounded by many species, new to them. But they did not seek out or describe those that held little value for food or clothing. They took little time to write about marine life in the bays and estuaries near where they wintered, including the estuary on which their winter home, Fort Clatsop, stood.

Their inspection and description of whales found along the coast did not increase when whale blubber was offered to them as food by the Indians. On January 5, 1806, the Indians gave two of the men "a considerable quantity of the blubber of a whale which perished on the coast some distance S. E.," Lewis wrote, adding that the blubber was "white & not unlike the fat of Poark, tho' the texture was more spongey and somewhat coarser." He ate it cooked and "found it very pallitable and tender, it resembled the beaver or the dog in flavour."

Two days later, Clark went to the coast where a team of men of the expedition maintained salt works, boiling sea water down to salt. He "hired an Indian to pilot me to the . . . whale." He mentioned meeting a group of Indians loaded with whale blubber, but made no other comment about this animal. On the same outing, he did describe "a Singular Species of fish which I had never before Seen one of the men Call this fish a Skaite."

Although the men of the expedition rarely ate shellfish, on December 9, Clark wrote that he was given dinner by the Clatsop Indians that include "Cockle Shells." On December 10, Clark saw Indians on the shore searching for fish washed up by the waves and tide, including "Sturgion."

Occasionally they fed on fish other than the salmon they had eaten so much of as they journeyed down the Columbia. The Indians gave them "Sturgion" on November 21, 1805. On that day, Clark refers to the food of the Chinnook Indians as "principally fish & roots the fish they precure from the river by the means of nets and gigs, and the Salmon which run up the Small branches together with what they collect drifted up on the Shores of the Sea coast near to where they live."

Even when Lewis reported "our meat is begining to become scarce," neither he nor Clark gave orders to seek whales or fish to be sought for food, let alone described. Perhaps because crossing and observing the land, not the ocean, was their primary task that they took little account of the abundant marine life. But when food supply was at its worst, just before the expedition began its return, Clark wrote on March 16, 1806, that there was "The pellucid jelly like Substance, called the *Sea nettle* I found in great abundance along the Strand where it has been thrown up by the waves and tide, and adheres to the Sand." And he saw two kinds of kelp, a large brown seaweed that forms a kind of marine "forest" important to many kinds of sea life on the Pacific coast, and an indicator that various edible fish and shellfish were present.

The California Condor

The same day that Clark saw the beached whale, he wrote that Ruben Fields killed a "Buzzard." This was the first time that anyone on the expedition had seen what we now call the California condor, and it is the first written observation of this species. Clark wrote the next day, once again measuring everything, that the distance from the tip of one wing to the tip of the other was "9 1/2 feet." The condor is the bird with the largest wingspan in North America. Modern range maps show the condor as restricted to southern California. A close relative of both the common American Turkey vulture or buzzard and the South American condor, the California condor is generally assumed to have survived primarily, if not exclusively, by feeding on carrion of deer, elk, and other land mammals. But later in the winter, on February 17, Lewis sketched the head of a California condor and wrote a long description. He wrote in characteristic detail about things he considered special and important, leaving the day-to-day journal entries to Clark.

"Shannon brought me one of the large carrion Crow or Buzzads of the Columbia which they had wounded and taken alive," he wrote, noting correctly, "I bleive this to be the largest bird of North America," measuring "9 feet 2 inches" from wingtip to wingtip. He then wrote about the natural history of the condor, first observing that "this bird flys very clumsily nor do I know whether it ever seizes it's prey alive, but am induced to beleive that it dose not." His observations of its feeding habitats run contrary to modern beliefs: "we have seen it feeding on the remains of the whale & other fish

Lewis's Drawing of a California Condor. *Lewis and Clark Codex J:80 (condor). Courtesy of American Philosophical Society.*

which have been thrown up by the waves on the sea coast these I beleive constitute their prinsipal food, but I have no doubt but they also feed on flesh." Adding circumstantial information about its food preferences, Lewis then wrote that "we did not met with this bird untill we had decended the Columbia below the great falls, and have found them more abundant below tide-water than above."

By the time the twentieth century opened, the California condor was in serious decline, numbering less than a hundred. To my knowledge, unlike Lewis no later writer uses the word "abundant" or the phrase "more abundant" in reference to the California condor: perhaps Lewis and Clark saw the bird when it was already undergoing a decline, but still common. Other observers suggest that Indians killed it for its great feathers, which might have contributed to its long decline. Lee Talbot, a twenty-first century expert on wildlife, ecology, and endangered species, has suggested, as did Lewis, in recent years that washed-up carcasses of marine mammals were likely to have been the dominant food of this bird, as Lewis suggested almost two hundred years ago. If so, then part of the condor's decline might have been the result of Yankee whaling and the decrease in numbers of grey and bowhead whales—the two inshore whales of the Pacific coast that were most likely to wash up on shore—during the great harvest. Its decline also could be tied to the decline in sea otters and the near-extinction of the elephant seal, which was hunted until only a dozen or so were left by the turn of the twentieth century.

What the California condor ate is more than a hobbyist's interest, as millions of dollars have been invested in an attempt to save this species from extinction and return it to the wild. By 1982, only twenty-two California condors were known to remain in the wild, and all of these lived in southern California. The state of California decided to remove the remaining condors from the wild and placed them in the Los Angeles and San Diego zoos, where captive-breeding programs began. These programs were successful in producing young condors, and the population is now more than 220, with 82 released and alive. The Peregrine Fund, which has taken over the condor program, has released birds into Arizona (where more than 30 are living), California (more than 40), and Baja California (less than 10). But current attempts to return the new adults to the wild—without the education such birds received about hunting and life in general

The California Condor in Flight Over a Condor Sanctuary. *U.S. Fish and Wildlife Service/David Clendenen.*

from their elders—have been fraught with difficulties, and the birds are still not able to find food for themselves.

Following on from the common assumption that California condors feed on carrion of land mammals, those involved in the reintroduction of the bird have done so in two areas: inland southern California and Arizona north of the uppermost Grand Canyon—dry areas with some, but not highly abundant, deer, elk, and mountain sheep. Some of the first birds let out of captivity died quickly, one from drinking antifreeze, another from being shot. And so those doing the reintroductions focused on habitats away from people. But little if any independent feeding has been observed. The birds feed primarily on stillborn calf fetuses trucked in and placed on ridges.

If this bird actually fed primarily on marine mammals, perhaps the efforts are misplaced. If California condors were brought to, say, the Channel Islands off the coast of Santa Barbara, they might be coaxed to feed on marine mammals carcasses. All marine mammals are protected from take, except under permits, by the 1973 Marine Protection Act. We might have to promote the killing of one listed endangered species to save another, a predicament officials have tried to avoid.

Wildlife Viewing Today near Cape Disappointment

Although the condor is gone from the Columbia River—gone everywhere in the wild except where it has been reintroduced by zoo breeding programs—there is still much wildlife to see along the Washington and Oregon coasts that would be recognized by Lewis and Clark. A visit to Leadbetter State Park, Washington, at the end of Long Beach peninsula, provides one such viewing place. Here one can see bald eagles and peregrine falcons, two species endangered in the second half of the twentieth century because of the use of DDT, which thinned eggshells of these and many other birds. These two species have recovered considerably, the eagle spectacularly; the peregrines, with only a few nesting pairs along Leadbetter beach, slowly.

Lewis and Clark not only came down the mighty Columbia River, they explored some of the smaller rivers along the Pacific coast. River scenery similar to what they experienced is probably best seen on the wild and scenic stretch of the famous Rogue River of Oregon, south of where Lewis and

The Rugged Oregon Coast Today, South of Fort Clatsop. This rugged country is like the landscape Lewis and Clark found along the Oregon coast. Changes from their time are apparent, especially the bridge across the river and the houses near the shore that appear as white rectangles. The cleared area below the airplane wing lacks the rectangular shapes of many twentieth century clearcuts. Snow on the distant mountains suggests how harsh the weather in this region can be in the winter, recalling how that winter challenged the Lewis and Clark expedition. Some of the grassy areas seen in the previous picture are in agricultural use, probably as pasture. Dairying is common today in this part of the Oregon coast. *D. B. Botkin.*

Clark traveled, but in similar topography, on similar bedrock, and in the same climate they experienced. The Rogue River is one of Oregon's most famous salmon-fishing rivers, and a long section of it, from near Grant's Pass not far from Medford, Oregon, to Paradise, a small resort about twenty miles from the coast, is protected under the Federal Wild and Scenic Rivers Act. As a result, this stretch of the Rogue is much used today for whitewater boating.

The Great Conifer Rain Forests of the Pacific Northwest

On November 19, 1805, Clark observed, "The hills next to the bay Cape disapointment to a Short distance up the Chinnook river is not verry high thickly Coverd. with different Species of pine &c. maney of which are large, I observed in maney places pine of 3 or 4 feet through growing on the bod-

The Rugged Rogue River. Rocks along the shore lie so steeply that it is difficult if not impossible to walk along that shore; river travel is the best way for people to pass. *D. B. Botkin.*

ies of large trees which had fallen down, and covered with moss and yet part Sound."

Then on December 8, Clark took "5 men" and "Set out to the Sea to find the nearest place & make a way, to prevent our men getting lost and find a place to make Salt, Steered S 620 W at 2 miles passed the head of a Brook running to the right, the lands good roleing much falling timber, lofty Pine of the Spruce kind, & Some fur, passed over a high hill & to a Creek which we kept down 11/2 miles and left it to our right, Saw fish in this Creek & Elk & Bear tracks on it, passed over a ridge to a low marshey bottom which we Crossed thro water & thick brush for 1/2 a mile to the Comencement of a Prarie which wavers, Covered with grass & Sackay Commis, at 1/2 Crossed a marsh 200 yds wide, boggey and arrived at a Creek which runs to the right." He saw elk, crossed the creek, went through "emence bogs, & over 4 Small Knobs in the bogs about 4 miles to the South & Killed an Elk, and formed a Camp." He wrote that "those bogs Shake, much Cramberry growing amongst the moss."

These observations are especially valuable to us because a major American environmental controversy of the latter part of the twentieth century, and the object of federal and state laws, regulations, and policies, is about the forests of the Pacific Northwest and their wildlife and fish. A common belief is that prior to European settlement, the Pacific Northwest coast was wall-to-wall old-growth forest (more than two hundred years old). A related belief is that this old growth is natural, and in fact the *only* truly natural state. Some scientists believe that before European colonization of the Pacific Northwest, a large fraction west of the Cascade crest (variously referred to as "the Northwest" or "the Douglas fir zone") was blanketed with dense evergreen forest, with 40 to 70 percent old growth or "late successional and old-growth forests" (older than eighty years).

Another related belief is that this old growth is the most important vegetation type for all biological diversity. Therefore, to save the biological diversity of the Pacific Northwest, as much of the land as possible must be put into old growth, or in the terms of those who believe that old growth is the only natural and desirable condition, *returned* to old growth. Large amounts of money have been spent on this issue, both in campaigns and in carrying out laws, policies, and regulations. These laws, policies, and regulations have had major economic and environmental effects in the Pacific Northwest. As a result much logging in this region has been stopped, leading timber corporations to increase harvest of forests in other nations, sometimes third-world nations that have fewer environmental regulations. So the effects have been worldwide.

Even if it were true that old growth is the only truly natural and good condition for the forests of the Pacific Northwest, the improvement in the ecological condition of this landscape are coming at the cost of the destruction of forests elsewhere in the world, unless, of course, there were international treaties that provided uniform standards of timber operations and forest management.

Therefore, it is important to know whether old growth really occupied most or all of the land in the Pacific Northwest coast at the time of Lewis and Clark and, if so, whether it was most beneficial to overall biological diversity, as well as specific species of plants, wildlife, and fish. Central to this question are the frequency, intensity, and effects of the two main environmental disturbances: fire and storms. And related to this is the role of Indians in creating forest and prairie fires.

During the winter stay at Fort Clatsop, Lewis had time to write and describe the major trees and a few of the major shrubs of the Oregon–Washington coast. He numbered them as he described them: (1) Sitka

spruce (never before given a written scientific description); (2) western
hemlock (which Lewis refers to as a fir); (3) grand fir (another species new
to science); (4) a fir not quite recognizable from his description, perhaps
grand fir again; (5) Douglas fir; and (6) western white pine (also new to science). Today western white pine is not found at the mouth of the Columbia
River.

Clark described Sitka spruce on February 4, 1806, writing that it "grows
to an emence size: verry commonly 27 feet in Surcumferonce at 6 feet above
the surface of the earth, and in Several instances . . . 36 feet in the Girth, or
12 feet Diameter perfectly Solid & entire." He wrote that they grew as tall as
230 feet and were often more than 120 feet high before branches started.

The now-famous Douglas fir, so much involved in controversies about
the conservation of old-growth forests, Lewis described in detail on February 6: "the bark thin, dark brown, much divided with small longitudinal

**Lewis's Sketch of a Douglas Fir Leaf,
February 6, 1806.** While at Fort Clatsop,
Lewis described what later was to be
named the Douglas fir, a famous and valuable timber tree of the Pacific Northwest.
*Lewis and Clark Codex J:65 (fir leaf). Cour-
tesy of American Philosophical Society.*

interstices and sometimes scaling off in thin rolling flakes. . . .twigs are much longer and more slender than in either of the other species. the leaves are acerose, 1/20th of an inch in width, and an inch in length, sessile, inserted on all sides of the bough, . . . and more thickly placed than in either of the other species; gibbous and flexeable but more stif than any except No. 1 and more blontly pointed than either of the other species; the upper disk has a small longitudinal channel and is of a deep green tho' not so glossy as the balsam fir, the under disk is of a pale green."

Anyone who has traveled the countryside of western Washington and western Oregon knows that the forests of the Pacific Northwest are impressive—huge trees tower over the land, darkening the ground, dominating everything, producing huge downed logs taller than a man—almost as tall as a horse—making travel almost impossible. So impressively large and dense are these forests that they seem permanent and unchanging. A hike into one of these ancient forests leaves us with the impression that the Pacific Northwest, prior to human exploitation, must have been a vast, almost mystical land of dark, giant forests extending over the entire landscape. But was that really the case?

Forests appear permanent to the casual visitor, in contrast to wildlife, which is often difficult to see and moves quickly away from human beings. To scientists, forests are much more open to study than wildlife, because trees don't move around and because in temperate and northern latitudes, trees lay down annual growth rings, which make it easy to determine their age. As a result, we know more about trees in forests and their interactions with each other and their environment than we do about wildlife.

Forests are an important part of our natural history, not only in terms of the resources and recreation they provide, but as symbols—they represent the idea of nature to us. Immense and seemingly boundless forests confronted the first settlers of the eastern United States—the ancestors of Lewis, Clark, and Jefferson, as well as the Pilgrims and their descendants. Clearing enough of those forests to allow room for crops to grow and towns to be built was a major early activity in North America. Forests have always provided essential resources in North America, just as they have for other civilizations. But compared to most Europeans, we who live in North America have much more direct contact with ancient forests, and the clearing of ancient forests over wide areas is much closer to us in time than it is in the modern European experience.

In recent years, forests have become a major focus of environmental controversy around the world. Much of the controversy has to do with the protection of ancient forests, which are steeped in symbolism. In part because of their appearance and the longevity of trees, forests uncut by human beings seem to symbolize the longevity and persistence of all life.

Old-Growth Forests and Native Americans

The contact between Lewis and Clark and the Indians along the Columbia River who used wood extensively and the experiences of the expedition in the forests of the Pacific Northwest lend valuable insight for us to apply today, when we debate the future of the forests of the Pacific Northwest.

Not many years ago, only a few citizens paid much attention to the forests, except to see them as places for occasional recreation, where few would venture far from well-marked trails. Most conservationists focused on individual species of large birds and mammals. But today, forests have become one of the major concerns among conservationists and loom large as a symbol of our natural history. These societal concerns are having an effect on state and federal agencies that manage forests. There is a shift from the traditional concern of these agencies on a single-factor goal of providing a maximum supply of timber to much broader goals under the name of "ecosystem management." This term and its companion "sustainability" have become the main jargon in the discussion of forests. Unfortunately, many meanings are attached to these terms.

To briefly characterize the controversy about Pacific Northwest forest, one side argues that use is good for forest and people, the other that use is good for neither. On one hand, timber companies and the U.S. Forest Service see these forests as major resources to be used, and private landowners, believing that they have fundamental property rights to the resources on the forest lands they own, resist government regulation. On the other hand, many members of environmental groups, as well as tourists traveling through these areas are shocked to see how large the clearcuts are, and how ugly they look—great rectangular barren slopes of brown rock and soil that appear abruptly next to some of the densest forests in the world. Whatever the real effect of these large clearcuts on the long-term persistence of the forests, this checkerboard effect gives a devastated look to the countryside, a look very different from the one seen by Lewis and Clark.

As stated earlier, the common belief about forests of the Pacific Northwest coast is that prior to European settlement, these were wall-to-wall old growth. To this I now add that this belief is consistent with the balance of nature that Jefferson was taught, and believed, and that influenced Lewis and Clark prior to the start of their journey. Implications of this view are that fire and storms did little damage to the forest; that forests could regenerate as old growth (that is, the species of trees that dominate an old-growth stand can also reproduce, persist, and take over dominance as the older individuals die within these forests); and that the Indians of the coast either did not light fires or, if they did, had little effect on the forest.

In a study of the relative effects of forest practices on salmon in the state of Oregon, conducted for the state legislature by the Center for the Study of the Environment, some historic maps were found in archives of the Bureau of Land Management. These maps help resolve a long-standing debate. They show forest conditions in 1850, 1890, 1920, and 1940, based on U.S. Geological Survey (USGS) measurements made on the ground, using methods that appear satisfactory to characterize vegetation conditions. These maps divide the landscape into five categories: forest 200 years or older; forest 100 to 199 years old; forest 50 to 100 years old; forest 0 to 49 years old; and recently burned areas. This indicates that forests of the Oregon Coast Range were not completely old growth. One-third was less than one hundred years old in 1850, and 34.5 percent was "recently burned" in the 1850 map. This is consistent with the idea that Native Americans frequently lit fires, and that a considerable fraction of Oregon presettlement landscape was subject to human-induced fires, as well as to some lightning-caused fires. With the settlement of Oregon by people of European descent, the suppression of fire began, and recently burned areas comprise only 5 to 7 percent of land in the three later maps.

What was the role of Indians in these forest conditions? As Robert Zybach, who has traced the history of the forests of the Oregon coasts, points out, the Indians of this region depended on many timber products and on many species of plants and animals, such as elk and deer, that require open prairie and young woodlands that would not have been found if the countryside were wall-to-wall old growth. Zybach also argues that since the Indians knew how to light fires, they were likely aware of the effects of fire on the landscape—for example, that fire allows the regeneration of many plants of direct use, including the major timber trees such as Douglas fir, and that fire was necessary for the game they hunted.

Zybach studies show that there were large areas of prairie and young forest, indicating that much of the area did burn. His evidence suggests that the Indians lit fires both intentionally, and indirectly as a byproduct of other activities. There are a few direct references to Indian-lit fires. One is an interview in 1991 with one of the older Indians, named Che-na-wah Weitch-ah-wah, who said, "The Douglas fir timber they say has always encroached on the open prairies and crowded out the other timber; therefore they have continuously burned it and have done all they could to keep it from covering the open lands. Our legends tell when they arrived in the Klamath River country there were thousands of acres of prairie lands, and with all the burning that they could do, the country has been growing up to timber more and more."

Zybach has developed a map of the vegetation of the Oregon coast near the Columbia River prior to European settlement, and therefore representative of the time of Lewis and Clark. This map only shows the larger open areas among the forests.

In 1845, a man named Wilkes wrote, "During the day they passed over some basaltic hills, and then descended to another plain, where the soil was a fine loam. The prairies were on fire across their path, and had without doubt been lighted by the Indians to distress our party. The fires were by no means violent, the flames passing but slowly over the ground, and being only a few inches high."

Zybach describes the houses of the coastal Indians, "The two upper pictures show a photograph of a plank house near the mouth of the Umpqua River taken in the 1850s; the drawing is of a similarly styled home, possibly Kusan, from the same time period. The lower right picture shows another 1841 drawing by Agate . . . of the interior of a Chinookan lodge. Consider the amount of firewood needed to heat structures of this size." One of these houses could require as much as 70,000 board feet of wood—the equivalent of two or three of even the largest Douglas firs. One near Portland, Oregon, 55 feet wide and 120 feet long, home to forty-five to sixty people,

Indians of the Pacific Northwest Coast, as they would have appeared to Lewis and Clark. *Bob Zybach*.

took a half to one million board feet just for maintenance and repair during its estimated four hundred years of use.

These Indians also built wooden canoes. "On January 20, 1806, Clark described this type of canoe in detail.

The . . . largest species of canoe we did not meet until we reached tidewater, near the grand rapids below, in which place they are found among all the nations, especially the Killamucks and others residing on the seacoast. They are upwards of 50 feet long, and will carry from 8,000 to 10,000 pounds' weight, or from 20 to 30 persons. Like all the canoes we have mentioned, they are cut out of a single trunk of a tree, which is generally white cedar, though the fir is sometimes used. . . . In this way they ride with perfect safety the highest waves, and venture without the least concern in seas where other boats or seamen could not live an instant. . . . In the management of these canoes the women are equally expert with the men.

Lewis and Clark visited many Indian villages, including those along the coast. On November 4, 1805, Clark described one such village, a Skilloot town, as "a village of 25 *Houses:* 24 of those houses we[re] thached with Straw, and covered with bark, the other House is built of boards in the form of those above, except that it is above ground and about 50 feet in length and covered with broad Split boards This village contains about 200 men of the *Skil-loot* nation I counted 52 canoes on the bank in front of this village maney of them verry large and raised in bow." This suggests the importance of timber for these coastal tribes.

Regarding the most recent Indian use of fire, Zybach writes that "a 1914 photograph taken of Soap Creek Valley, [shows] that a combination of farming, grazing, and slashing has somewhat retained the 'cultural legacy' of earlier generations of Kalapuyans . . . . However, a 1989 picture from the same perspective paints an entirely different picture. Farming practices that replaced burning practices have themselves been replaced with forestry practices. The oak savannah of the Kalapuyans and the open grazing lands and fenced pastures of early white settlers have been replaced by Oregon State University Research Forests' timber: the McDonald and Dunn forests."

Lewis and Clark achieved one of the major goals of their expedition: They traveled from St. Louis to the mouth of the Columbia River. An uncelebrated but important result of this journey is the knowledge forced upon them by the land and rivers they passed through. They came to know a nature unlike that depicted by Jefferson's educators, unlike that represented in the traditions of western civilization. They traveled two of the great rivers of the world, the Missouri and the Columbia, and found that these changin' old rivers symbolized the character of nature in the American West. It was a nature in which change was natural, and within which people had to understand, accept, and make use of change in nature if they were to survive and persist. It is a lesson that we—and even the descendants of the coastal indians—have too often forgotten.

14

IN THE WAKE OF LEWIS AND CLARK

THE WINTER AT Fort Clatsop had been cloudy and rainy, with few days of any sunshine. Food ran low. The men were often sick. But Lewis wrote in his journal on March 20, 1806, the day before they were to leave the Pacific coast, "Altho' we have not fared sumptuously this winter and spring at Fort Clatsop, we have lived quite as comfortably as we had any reason to expect we should; and have accomplished every object which induced our remaining at this place except that of meeting with the traders who visit the entrance of this river."

From our modern perspective, this is an amazingly positive comment about what must have been a dismal winter. I have visited the restoration of Fort Clatsop on Christmas Eve, when the National Park Service holds a holiday ceremony. At that time of year, it is likely to be cloudy, raining (or drizzling), and cold. The air is damp, and even the middle of the day feels dark. One's bones chill. Sitting in the reconstructed fort, even for a short ceremony, is a miserable experience. Yet Lewis and Clark and their crew spent months there, many of the men ill with colds or flu at the outset.

Was Lewis's entry bravado, or did he really believe it? Did he not have moments when he wondered whether it had all been worth it? He never expresses doubt in his journal. But now that we have followed Lewis and Clark's journey across the continent from the perspective of nature in the American West, we may wonder: Was it worth it? What does the journey mean? What did it accomplish?

At the least, the journals help us understand what has happened to nature in the American West since European settlement. Perhaps most startling is that the least-celebrated ecological regions—the tall- and short-grass prairies—are the most changed, difficult to find even as small remnants. Of the two major rivers, the Missouri—the one of less concern, both to scientists and to environmentalists—is more altered than the Columbia. Of the ecological regions visited, the one that is most like what Lewis and Clark saw—the Bitterroot Mountains—is much less in the spotlight than any other they visited. If of value in no other way, the journals are a unique perspective on how nature has changed.

Lewis and Clark's journey also gives insight about the *idea* of nature. They found a nature much different from the European expectation: they found a nature of constant change, in which what would happen could at best be estimated as probabilities. It wasn't possible to ensure that everything would be safe. Like Lewis and Clark making their decision at the mouth of the Marias River, we sometimes have to make do with the best that we know. Sometimes chance events, like Lewis's slipping on mud on his way back down the Marias, impose themselves. In our attempts to control nature since Lewis and Clark, we have learned that sometimes the harder you try to make everything safe, the worse you fail. Nothing illustrates this better than the attempts since 1806 to control the Missouri River and to make passage on it safe, with neither flooding nor drought. Our relationship with nature, and our view of it, must accept its unpredictability for the good and the bad. But our technological prowess tends to lead us in the other direction: The greater our apparent control over nature, as with the Missouri River, the greater faith we have in our own effectiveness and the less alert we are to possible dangers.

Lewis and Clark set out into the American West during the rise of science and technology. We have since come to believe we could control nature as if it were a machine we made ourselves—a horse-drawn carriage, a steamboat, a hydroelectric generator. But that belief doesn't seem to solve problems when we try to treat nature that way—not on the Missouri River and not on the Columbia.

At the Spirit Mound, Clark demonstrated an exceptional ability to observe, think, analyze, and later write down his thoughts. Such abilities are uncommon in our age—few of us are used to a life within nature in the way that Lewis and Clark experienced it. We might be, like Lewis and several of the men, desperately thirsty to the point where we can think of little else, but Clark carries on, measuring, reflecting, and writing. It is a kind of heroism we are not accustomed to celebrating, nor even used to imagining.

At the Great Falls on the Missouri River, and at many locations where Lewis and, more rarely, Clark stepped out of their role as only scientific reporters, they found the landscape incredibly beautiful. At the confluence of the Yellowstone and the Missouri, they found an American Serengeti, with wildlife as tame as in old legends.

So at both the external and internal levels of our experience of nature, Lewis and Clark provide insight.

There is one more major change that has taken place since the beginning of the nineteenth century, which we have not dwelt on yet: America has become an urbanized society, with about three-fourths of us living in cities or suburbs. This has altered the landscape, and it has altered our perception of nature. When one thinks about biological "nature," one usually thinks about wilderness, nature preserves, national parks, wildlife, endangered and rare species—a "natural" landscape, in one sense or another. But a consideration of nature then and now would be remiss if it did not recognize that large areas of the landscape in the American West have been converted from prairies, forests, and wild rivers to agriculture, cities, and towns. We have discussed the loss of the prairies and forests elsewhere in this book. This last chapter will focus on the urbanization of the landscape in the wake of Lewis and Clark.

The future of our environment lies as much in the way we treat our cities as it does in the way we treat our wildernesses, perhaps even more so. The more pleasant our cities are to live in, the more people will want to live in them. The more people want to live in cities, the less pressure people will place on outlying areas for development and the more likely that there will be room to share the landscape with other creatures. Wilderness, wildlife, prairies, and forests will benefit if our cities prosper.

The major cities of the Lewis and Clark trail are, on the Missouri River, St. Louis, Kansas City, and Omaha; and on the Columbia River, Portland, Oregon. Each differs in the connectedness between the city and its river, and this has had a profound effect on the success of each city—success in terms of how livable the city is today, how vibrant and how economically productive.

St. Louis, Missouri

On May 14, 1804, the expedition left Camp Dubois for St. Charles, Missouri. The town was of some importance at that time, a short distance up the Missouri—the first white settlement north of the Missouri and west of the Mississippi, and today a historic suburb of St. Louis. Clark took the boats and the crew while Lewis walked so that he could finish some business in St. Louis.

Clark with the crew and boats arrived at St. Charles on May 16 and observed the town carefully, as he would observe all landscapes on the trip. St. Charles contained about one hundred houses, he wrote, most "Small and indifferent." On the way, they passed Coal Hill, which he understood "Contain great quantity of Coal & ore."

Meanwhile, Lewis carefully observed the countryside as he walked from Camp Dubois to St. Charles. On May 20, he wrote, "The first 5 miles of our rout laid through a beatifull high leavel and fertile prarie which incircles the town of St. Louis from N. W. to S. E."—land that today would be well within the urban development of that city.

St. Louis was founded and designed by Pierre Laclède, a Frenchman who arrived there in 1763 to build a fur-trading post at the confluence of the two rivers. The post had to be above the rivers' flood levels, and he wanted his town to become "one of the finest cities in America." He surveyed the area and found that there was a high limestone bluff forty feet above the river and therefore safe from floods. This is now the location of downtown St. Louis.

The confluence also had the advantage of proximity to forests: at that time, as throughout most of the history of civilization, timber was an essential commodity, for buildings, furniture, tools, and fuel. The town began quite small: in 1766, it had only about seventy five buildings and three hundred residents. By the time Lewis and Clark arrived, the number of houses had more than doubled, and the population was about nine hundred.

St. Louis and Its Famous Arch. The arch was part of one of several attempts by the city to restore the downtown. However, the city itself is separated from the Missouri River by a highway, which in turn runs along a channelized bank. The arch and its park function more as an island between a river and a decaying downtown than as a path bringing people to the river and the river to the people. *U.S. Army Corps of Engineers.*

How well has St. Louis fared in terms of Laclède's original vision? This city has made three attempts to renovate its center and is in the midst of a fourth. The most famous of these attempts was the construction of a low-cost housing project that was so unsuccessful that the buildings were dynamited away. The famous arch is beautiful, but St. Louis has not succeeded in becoming the focus of a major urban renewal. Why not? At least part of the answer lies with the connection between a city and its river.

Kansas City, Kansas

The site of Kansas City also appealed to Lewis and Clark as described in chapter four. On June 25, 1804, the expedition camped in what is now Sugar Creek, a suburb of Kansas City. The expedition had entered the second ecological region, the tall-grass prairie, leaving behind them the eastern deciduous forest of Missouri and most of the states east of the Mississippi. They spent several days there because they had to repair the pirogue, which they emptied, brought up on land, and turned over. During this work, on June 28, the expedition saw its first buffalo, which it did not kill. These animals would soon become a principal source of food.

That day, the expedition camped just above the mouth of the Kansas River, at the present location of Kansas City, Kansas. It is worth repeating that Clark praised the location for its beauty and its opportunity for defense. He wrote that the Kansas Indians lived "in a open & butifull plain" and that "the high lands Coms to the river Kanses on the upper Side at about a mile," which made "a butifull place for a fort, good landing place."

Kansas City began as a trading post in 1821, only fifteen years after Lewis and Clark returned past the mouth of the Kansas River on their way back, and only two years after the first steamboat sailed on the Missouri. The town grew as steamboat activity increased. Other towns and cities along the Missouri developed with similar speed and without the late-twentieth-century concern for the environment that we take for granted.

Kansas City was an important location for a geographic reason: Downstream, the Missouri runs east and west; above the Kansas River, it runs north and south. This was the farthest west one could go on the Missouri below Sioux City, where the river once again turned west. As explained in chapter four, a traveler at the Kansas River mouth could choose to take that smaller river, which few did; or go north to Omaha and take the Platte River west, which became the major route west; follow Lewis and Clark and continue up the Missouri; or begin travel by land. This made the location a natural one for a city as well as a fort.

And how has this city fared? Within the city, the Kansas River flows through an underground aqueduct, so it is no longer a part of the landscape. It is difficult to get near the Missouri River. You can get a view of the river near the confluence at Riverfront Park in Kansas City, which can be reached from Interstate 29/35 Front Street southbound ramps. But the city airport lies on the river's north shore, and there are roads, railroads, and factories, some in decay, along the south shore. Although Kansas City is proud of its trees and its planning, it is not well connected to its two rivers and not recognized as a major, important city elsewhere in the nation.

Kansas City Seen from Space. A view toward the east showing the confluence of the Kansas River, flowing east, and the Missouri River, flowing here southeast. Visible are agriculture and urban areas. After Lewis and Clark, this location was favored as the starting point of expeditions and wagon trains of settlers. The confluence of the two rivers, and the location where the Missouri River turns from south-flowing to southeast, makes this as far west as one could go directly by major waterway from St. Louis in solely a westward direction. *Image courtesy of Earth Sciences and Image Analysis Laboratory, NASA Johnson Space Center. STS040-88-00H, http://eol.jsc.nasa.gov*

Fort Benton and the Geography of Cities

The importance of the geographic situation of a town or city, and of rivers to most cities is illustrated by Fort Benton, Montana. When Lewis and Clark approached the Rocky Mountains, they realized that it would be necessary to leave some of their equipment behind. They had to lighten their load as much as possible to get over the mountains, and there were some things, such as their boats, that they could not bring up the Rockies. They had to cache their heavy equipment, and the location they chose was near what is now called Fort Benton, Montana.

Downstream a little way from this town, they dug a large hole, like a house basement. To be on the safe side, they stored some gunpowder and lead—"To guard against accedents," they noted in the journals—in case they lost the rest and needed more on their return. They also left two of their "best falling axes, one auger, a set of plains, some files, blacksmiths bellowses and hammers Stake tongs &c. 1 Keg of flour, 2 kegs of parched meal, 2 kegs of Pork, 1 Keg of salt, some chissels, a cooper's howel, some tin cups, 2 Musquest, 3 brown bear skins, beaver skins, horns of the bighorned animal, a part of the men's robes clothing and all their superflous baggage of every discription, and beaver traps." They tied down a boat, their red pirogue, on a small island in the river and covered it with brush.

The first steamboat to navigate the Missouri, the *Independence*, moved up her waters on May 28, 1819, only twelve years after Fulton's steamboat sailed on the Hudson River and only thirteen years after Lewis and Clark returned to St. Louis. As the Pacific coast opened up and people sought better ways to travel west, steamboats began to take people and materials up the Missouri. In 1846, a town was established near the site where Lewis and Clark cached their equipment. First called Fort Lewis but renamed in 1850 for Senator Thomas Hart Benton of Missouri, the town became the terminus of steamboat travel. Fort Benton boomed during the California Gold Rush as people rushed to get to the West coast and as cattlemen began to use steamboat transportation for supplies. From here, travelers took the Mullan Wagon Road, 624 miles from Fort Benton to the beginning of navigable waters on the Columbia River, and for years it was the fastest route, taking forty-seven days.

Why did this become a common place to halt and either cache excess baggage or stop bringing large boats farther? Why did Lewis and Clark not wait until they had reached the Great Falls, the truly impassable section of the river, and cache their equipment just below that? Or why not leave things farther downstream than the area near Fort Benton?

For both the expedition and for steamboats, stopping many miles downstream from Fort Benton would have been difficult. At the site of modern Fort Peck Dam, the Missouri River begins its passage through steep bluffs and cliffs, and these continue to Virgelle, Montana. Even if Lewis and Clark had found a place to cache their goods in that section of the river—a place that would have been safe from flooding and where the soil was deep enough to store things—they would have had a difficult time finding a good trail that led down to the river on a gentle slope.

Knowing that the Rocky Mountains could not be too distant, it would have been a natural decision for explorers to stop and cache their goods as soon as the land began to flatten out again. This is what happens near Fort Benton. Thus the geology of this location made it a good place to take things *to*, up the Missouri, but not take things *beyond*. Lewis and Clark, as well as later travelers going west, were once again affected by the geology and the geological history of the Missouri River basin.

So it is with most major cities, though travelers are often unaware of it. Most major cities around the world lie at crucial locations along rivers, one of three kinds of locations. The first is the ocean mouth of a river, as with New York City and New Orleans. The second is the junction of two major rivers—the site of St. Louis, where the Mississippi and Missouri come together, the site of Omaha, Nebraska, where the Platte River flows into the Missouri, and Portland Oregon where the Willamette River joins the Columbia. The third is at what is called the "fall line"—where a river on its way downstream passes from harder, more erosion-resistant rocks to softer rocks, resulting in waterfalls or unnavigable rapids.

The fall line is a natural location to create a city and a natural place for a city to succeed. The fall line is not only the farthest inland that a steamboat or ship can navigate, but is also typically far enough upstream to be easily spanned by a wooden bridge, important before the invention of modern steel-suspension bridges. And the falls are a good site for waterpower. Great Falls, Montana, is just upstream of a fall line; Fort Benton is well situated not far below it.

Usually, a fall line is relatively near the ocean—within a few hundred miles. This is the case with many major cities of the East. In Jefferson's Virginia, the city of Richmond is on a fall line, as are most of the inland cities of the East Coast and south-central plains, including San Antonio and Fort Worth, Texas; Little Rock, Arkansas; Montgomery, Alabama; Columbia, South Carolina; Washington, D.C.; Baltimore, Maryland; and Philadelphia, Pennsylvania. On the Missouri, the only odd thing is that the fall line occurs a longer distance from the ocean, at Great Falls. Thus, with the building of the railroads, Fort Benton diminished in importance as a transportation terminus and transit point. It remains a pleasant small town—one of the nicest places to view the upper Missouri in Montana.

Portland, Oregon

On April 2, 1806, the expedition was on its way back, still on the Columbia River, when Lewis and Clark found a large river flowing into the Columbia from the south that they had missed on the downstream trip the previous fall. The river, today known as the Willamette, was called "*Mult no mah*" by local Indians who lived on Wappato Island, a little below the mouth of this river. Clark wrote, "The Current of the Multnomar is as jentle as that of the Columbia glides Smoothly with an eavin surface, and appears to be Sufficiently deep for the largest Ship." He tried to measure the depth but had only a rope of five fathoms—30 feet—and with this he could not find the bottom. Later he wrote, "I think the wedth of the river may be Stated at 500 yards and Sufficiently deep for a Man of War or Ship of any burthen."

Clark notes that they had missed this large tributary on the way down because it had been obscured by islands as they passed it. On the return, he went up this river by canoe for "10 miles." He was right about the capacity of this river. Its confluence with the Columbia would become Portland, Oregon, the major ocean port of the Columbia River, even though it was, according to Clark's measure on April 13, "142 miles up the Columbia river from its enterance into the Pacific Ocean."

Usually the major ocean port of a large river is at the river's mouth. At the mouth of the Columbia, John Jacob Astor established a settlement in 1811, only five years after Lewis and Clark left Fort Clatsop. He did so to promote fur trade with the northwest Indians, and one would have expected that settlement, Astoria, to become the Columbia's major port. But it did not. It is a port, but a minor one compared with Portland. The reason Portland became the Columbia's major port is explained by an observation that Clark made on April 5, 1806. "The Country a fiew miles up the Multnomah River rises from the river bottoms to the hight of from 2 to 300 feet and is rich & fertile." This was the north end of the great Willamette valley, later to become a major agricultural valley. It was much easier to transport the products of that agriculture, including grain and meat to the ocean, by boat than overland, especially over a trail that required crossing the coastal mountains of Oregon. Although not high compared with the Rockies, these mountains have a complex topography of many steep hills and valleys that create a difficult passage; this explains the establishment of Portland and its rise as a city.

The original function of Portland was somewhat similar to that of St. Louis, and to a lesser extent Kansas City. Other things being equal, one might expect these cities of the Lewis and Clark trail to fare similarly into the twentieth century. But Portland is today a vibrant, vital city, the Pacific Northwest's center for jazz; a city with excellent live theaters. The city center, in contrast to St. Louis and Kansas City, is integrated into the landscape design and is a pedestrian's joy. The difference between these cities seems to be partly the result of how the citizens and their leaders approached urban life and restoration.

In the mid-twentieth century, the construction of Interstate 5, the highway passing along the western shore of the Willamette, separated central downtown Portland from the river. Similarly, the center of St. Louis was separated from the Missouri River by a major highway. Slums developed along the St. Louis river shore. The highway now runs in a kind of open tunnel, below the level of the other city streets and sidewalks. One might think this would allow the city and the river to be better connected than they were, but the noise, exhaust, and general appearance of the highway act as barriers. St. Louis has made three attempts to restore its downtown. But the movers and shakers of the city do not choose to live there, and the restoration does not succeed in connecting the river to the city.

In contrast, the city of Portland raised funds to move Interstate 5 across the Willamette River, to the east side, and replaced it with Riverside Park, with easy foot access to the city center. Small parks abound in Portland. Some pedestrian walkways pass in the middle of blocks rather than along the street, and these have small parks—a few benches, perhaps a sculpture, a fountain, or other decorations. The west of the city is occupied by the largest forested park in the world, Forest Park, running much of the length of the city on a steep slope that rises above downtown. The firm of Frederick Law Olmsted, the great father of park planning who designed Central Park in New York City and many other major parks in America, proposed a design for Forest Park, a design that has attracted good attention recently. It is called the largest forest wilderness park in the world.

Olmsted believed that parks play an important role in cities. He wrote that vegetation in cities provides medical, social, and psychological functions. The Olmsted plan for the park stated that "no use to which the land could be put would begin to be as sensible or as profitable to the city as that of making it a public park or reservation." His plan languished until after

Forest Park, Portland Oregon Called "the Largest Urban Wilderness Park in America." The park covers hills that overlook the city and the Willamette River. Established in 1947, Forest Park occupies 5,000 acres and includes a zoo, the World Forestry Center, a rose garden, and a Japanese garden. *Courtesy of Friends of Forest Park.*

World War II, when citizens of Portland became interested in it and made it a reality. The park was established in 1948. Among its many benefits to the city, Forest Park has provided sightings of more than 110 species of birds and 50 species of mammals.

Although many factors determine whether a city will persist and prevail as a major center for commerce and creativity, it has long been recognized—for more than two thousand years—that one of the main functions of city planning is to create urban beauty. Why? The belief is that the more beautiful a city appears, the more it will attract people to live in it, and the more likely it will to succeed both in commerce and as a center of civilization. We tended to ignore this in the twentieth century, since railroads, cars and highways, and airplanes and airports appeared to make us independent of geographic constraints. But we are reminded of this by the present conditions and the histories of the cities along the Lewis and Clark trail—St. Louis, Kansas City, and Portland. When we compare these, the argument can be made that still today the better a city is connected to its river—in ease of access and in landscape beauty and design—the better the city functions. We make cities more beautiful by bringing nature into them. This re-attracts people who left to seek a bit of green in the suburbs. So those interested in conserving wilderness and those interested in the quality of urban life share an important goal—to concentrate human populations in attractive urban centers and to free more of the landscape for forests, prairies, rivers, wildlife, fish, and beautiful scenic landscapes.

What lies in the wake of Lewis and Clark? Our cities and their futures. Our ability to understand nature and benefit from that understanding. Our connection between people and nature. And in the broadest terms, ourselves, inward and outward, as we live within this world and are connected to nature as Lewis and Clark saw, recorded, and understood it.

Source Notes

All quotations attributed to the Lewis and Clark journals, unless otherwise indicated, are from *The Journals of Lewis and Clark,* Gary Moulton, ed., vols. 2-8 (Lincoln: University of Nebraska Press, 1986-). This is the best, and now only authoritative edition of the journals, excellent in its careful and useful notes as well as the editing of the entries themselves. Should a reader want to find a quotation in the journals, they can do so with the entry date, by which the Moulton edition is organized. Due to the ease with which the quoted entries can be located, and in an effort not to clutter the text, the quotations have not been referenced in the text. If the date alone is not sufficient to locate a quotation, a citation will appear in this section, located in the text by page number and the first few words of the quotation, with editor (Moulton), volume number: page or page range indicated.

Chapter 1: A Partially Settled Landscape

Page 1   "By two experiments. . . ." Moulton, II: 169–170.

Page 2   The thirty-three members of the expedition . . . . An excellent discussion about the members of the expedition can be found in Moulton, II: 509–529 (Appendix A). For the purposes of this book, the following summary is helpful. The expedition included Lewis and Clark, nine men from Kentucky, fourteen U.S. Soldiers, two French watermen (Cruzatte and Labiche), an interpreter and hunter (Droulliard/Drewyer), a black servant (York). All except York were appointed as privates in the Army; three were then made sergeants (Floyd, Ordway, Pryor). Later, with the death of Floyd, Patrick Gass, a private, was promoted to sergeant. In addition, a corporal, six soldiers, and nine watermen were employed to accompany the expedition to the Mandan Indians. Thus there were 45 people, of which 16 were engaged only to go as far as the Mandan Indians; the names of these 16 are unknown except for a Corporal Warfington. The Canadian, Toussaint Charbonneau, who served as an interpreter, his Indian wife, Sacagawea, and her baby boy, Jean Baptiste, joined the expedition at the Mandan Villages. Only one man, Charles Floyd, died on the expedition.

Journals were kept by Lewis, Clark, and by each of the sergeants, the most notable of which is that by Sergeant Patrick Gass. His journal was published first in 1807 (before the publication of the journals of Lewis and Clark); Gass's journal is included in the Moulton edition of the journals as volume 10.

Page 3   "Record the mineral productions of every kind. . . ." The letter from Thomas Jefferson to Lewis is quoted from Elliott Coues, ed. *History of the Expedition under the Command of Lewis and Clark,* 3 vols. (1893; reprint, New York: Dover Publications, 1965), 1: xxvi. Subsequent references to this source will be cited as follow: Coues, volume number: page range.

Page 11   One of the best maps of the late eighteenth century. . . . From the University of Virginia Library, Albert and Shirley Hall Special Collections Library, Lewis and Clark the Maps of Exploration 1507–1814. Available at www.lib.virginia.edu/speccol/exhibits/lewis_clark/planning3.html (last accessed April 14, 2004). This website contains the following statement: "Aaron Arrowsmith was considered the finest mapmaker of his day. He produced his first map of North America in 1795 from data collected from the archives of the Hudson's Bay Company. Arrowsmith's 1795 map shows a vestige of the 'Great River of the West' and the Missouri River appears as a river fragment unconnected to either the single ridge of the 'Stony Mountains' or the Mississippi River. Arrowsmith notes that the 'Stony Mountains' are '3250 Feet High Above the Level of their Base and according to the Indian account is five Ridges in some parts.'"

Page 11   Jefferson was taught by a Reverend Maury. . . . Jefferson wrote in his autobiography that "on the death of my father I went to the revd Mr. Maury a correct classical scholar, with whom I continued two years, and then went to Wm. and Mary college, to wit in the spring of 1760, where I continued 2. years." The Avalon Project at Yale Law School, *Autobiography by Thomas Jefferson, 1743–1790.* Available at www.yale.edu/lawweb/avalon/jeffauto.htm (last accessed April 12, 2004).

Page 13   "Accept no soft-palmed gentlemen. . . ." R. Dillon, *Meriwether Lewis: A Biography* (Coward-McCann, N.Y.: 1965), p. 58.

Chapter 2: Changing Old River

Page 18   "at half passed three. . . ." Moulton, II: 44.

Page 20   "Some people would think. . . ." C. D. Steward, "A Race on the Missouri," *The Century Magazine,* LEX no.4 (February, 1907): 588.

Page 29   "Next, the Army Corps of Engineers began a major. . . ." The information about the history of modifying the Missouri River is from U.S. Army Corps of Engineers, Northwestern Division August 2001, *Missouri River Master*

Water Control Manual, "Review and Update, Revised Draft, Environmental Impact Statement," vol.1: Main Report, *Contents Summary*, (August 2001).

Chapter 3: A Countryside Pleasant, Rich, and Partially Settled

Page 47   "when these rivers [Missouri and its tributaries] form new lands on their borders or Islands in their steams . . . ." Moulton, II: 452. Lewis's Fort Mandan Miscellany, undated, but written during the winter of 1804–1805.

Page 48   "As the willow increases in size. . . ." Moulton, II: 452. Lewis's Fort Mandan Miscellany, undated, but written during the winter of 1804–1805.

Page 50   "The Kickapoo calls a certain. . . ." Moulton, II: 221–223 (Lewis, undated journal entry).

Page 51   "The surface of the cone. . . ." Moulton, II: 221–223 (Lewis, undated journal entry).

Page 52   "[T]his they boil untill . . . ." Moulton, II: 221–223 (Lewis, undated journal entry).

Page 53   "called by the Chipeways Moc-cuppin . . . ." Moulton, II: 224 (Lewis, undated journal entry).

Chapter 4: Into the Tall-Grass Prairie

Page 59   "Another early prairie traveler, Josaih Gregg. . . ." D. B. Botkin, "The vegetation of the West," in H.R. Lamar, ed., *The Reader's Encyclopedia of the American West* (Thomas Y. Crowell: N.Y., 1977), pp. 1216–1224.

Page 71   "General Isaac I. Stevens, who on July 10, 1853, was surveying. . . ." F. Haines, *The Buffalo* (Thomas Y. Crowell: N.Y., 1970), p. 33.

Page 71   One of the better attempts to estimate the number of bison in a herd was by Colonel R. I. Dodge . . . . M. Sandoz, *The Buffalo Hunters* (Lincoln: University of Nebraska Press, 1954), p. 102.

Page 72   "a train traveled one hundred. . . ." Sandoz, p. 83.

Page 73   "a great store of cattle." Samuel Argoll is quoted by Haines, p. 73.

Chapter 5: Restoring the Lower Missouri River

Page 92   "I like to bring people up here and show them . . . ." J. C. Bryant, former director of the Big Muddy National Wildlife Refuge, personal communication, 1998.

Chapter 6: Fire, Wind, and Water

Page 94   The loess hills and their prairies are based on material from the State of Iowa Department of Natural Resources publication, *IOWA—Portrait of the Land—A Century of Change: 1800 to 1900*, available at http://www.igsb.uiowa.edu/portrait/3change/change.htm (last accessed April 14, 2004).

Chapter 7: Lewis and Clark Among the Mandans

Page 125   "The black-tailed prairie dog. . . ." The current and past status of the prairie dog is available at http://www.state.sd.us/doa/prairiedog.htm (last accessed April 14, 2004).

Page 125   "the decline of the black-footed ferret. . . ." The current and past status of the black-footed ferret is available at www.ngpc.state.ne.us/wildlife/ferret.asp (last accessed April 14, 2004).

Chapter 8: America's Serengeti

Page 144   "Laliberte and Ripple. . . ." Their estimation of the number of grizzlies seen by Lewis and Clark and their estimate of the 1804 abundance of these animals is published in Andrea S. Laliberte and William J. Ripple, "Wildlife Encounters by Lewis & Clark: A Spatial Analysis of Interactions between Native American and Wildlife," *BioScience,* vol. 53 no.10 (2003): 994–1003.

Page 146   "to establish viable, self-sustaining populations. . . ." Information about the Grizzly Recovery Plan is available at http://mountain-prairie.fws.gov/endspp/grizzly/ (last accessed April 14, 2004).

Page 148   See Paul Martin, *The Last 10,000 Years* (Tucson: University of Arizona Press, 1963); and P. S. Martin and C. R. Szuter, "War Zones and Game Sinks in Lewis and Clark's West," *Conservation Biology,* vol. 13, no. 1 (1999): 36–45.

Page 150   "In 1867, the estimates of the number of bison between the Platte River and the Concho River in Texas. . . ." This and other nineteenth-century accounts are from M. Sandoz, *The Buffalo Hunters* (Lincoln: University of Nebraska Press, 1954).

Chapter 9: Scenes of Visionary Enchantment

Page 157   "Average maximum temperatures in the summer during the past 30 years. . . ." National Weather Service Forest Office, Great Falls, Mont., "What is Normal?" Available at http://www.wrh.noaa.gov/Greatfalls/misc/NORMALS.pdf (last accessed April 14, 2004).

Chapter 10: "Pleasingly Beautiful" and "Sublimely Grand"

Page 173   "The Montana Land Reliance has obtained conservation easements . . . ." Available at http://www.mtlandreliance.org/waters.htm (last accessed April 14, 2004).

Page 175   "It's not just the continents that move. . . ." Brian Skinner, S. Porter, and
            D. B. Botkin, *The Blue Planet* (Hoboken, N.J.: John Wiley: 1999).

Chapter 11: A Passage Steep and Stoney, Strewn with Fallen Timber

Page 193   "Opened a pack on his horse . . . ." The accounts of the Indians in the Bit-
            terroot Mountains about their meetings with Lewis and Clark are from Z.
            L. Swayne, *Do Them No Harm!* (Legacy House: Orofino, Idaho, 1990).
Page 194   "Wa-kuese . . . ." Swayne, *Do Them No Harm!*

Chapter 12: Roll On, Columbia, Roll On

Page 207   "swarm of large basalt dikes. . . ." Information about the geology of the
            Columbia River Gorge and the Pacific Northwest are based on D. D. Alt,
            and D. W. Hyndman, *Roadside Geology of Oregon* (Mountain Press: Mis-
            soula, Mont., 1978), p. 166. This and other "Roadside Geology of" books are
            among the best references for the traveler about the geology of the amazing
            route of Lewis and Clark.
Page 217   "The total catch of chinook, coho, sockeye, chum, and steelhead. . . ." and
            other historic information about salmon catch is from V. W. Kaczynski and
            J. F. Palmisano, *A Review of the Management and Environmental Factors
            Responsible for the Decline and Lack of Recovery of Oregon's Wild Anadro-
            mous Salmonids*, Technical Report (Oregon Forest Industries Council:
            Portland, Oreg., 1992): 202–204.
Page 222   "From the beginning, Bonneville Dam has symbolized hope. . . ." U.S.
            Army Corps of Engineers, *Bonneville Lock and Dam in Celebration of our
            50th Year*. Available at http://www2.kenyon.edu/projects/dams/bonne.html
            (last accessed April 14, 2004).
Page 223   Woody Guthrie lyrics are from B. Murlin, ed., *Woody Guthrie Roll On
            Columbia: The Columbia River Collection* (Sing Out: Bethlehem, Pa., 1991),
            p. 40. Used with permission of The Woody Guthrie Foundation and
            Archives, 250 West 57th Street, Suite 1218, New York, N.Y., 10107.

Chapter 13: Changing Old Forests at the Mouth of the Columbia

Pages   "the dates at which particular plants . . . . " Coues, I: xxvi, letter by Jefferson
230–31   regarding Lewis.
Page 242   Salmon status and trends is from D. B. Botkin, K. Cummins, T. Dunne, H.
            Regier, M. J. Sobel, and L. M. Talbot, *Status and Future of Salmon of Western
            Oregon and Northern California: Findings and Options* (Center for the
            Study of the Environment: Santa Barbara, Calif., 1995).
Page 243   "The Douglas fir timber. . . ." Robert Zybach, "The Great Fires: Indian

Burning and Catastrophic Forest Fire Patterns of the Oregon Coast Range, 1491–1951"(Ph.D. diss., Oregon State University, 2003).

Page 245   "The largest species of canoe. . . ." Zybach, Ph.D. diss., 2003.

Page 246   "and 1914 photograph . . . ." Zybach, Ph.D. diss., 2003.

Chapter 14: In the Wake of Louis and Clark

Page 250   "one of the finest cities in America" Available at http://www.stlouiswalkoffame.org/inductees/pierre-laclede.html (last accessed April 14, 2004).

Page 259   "no use to which the land could be put . . ." Hornburg I. Wagman, *St. Louis History*, "History of St. Louis Neigborhoods," available at http://stlouis. missouri.org/neighborhoods/history/ (last accessed April 14, 2004).

INDEX

* Note: Page numbers in **bold** indicate chapter ranges. Page numbers in *italics* indicate photographs and illustrations.